Artificial Intelligence and Blockchain in Digital Forensics

RIVER PUBLISHERS SERIES IN DIGITAL SECURITY AND FORENSICS

Series Editors

The "River Publishers Series in Security and Digital Forensics" is a series of comprehensive academic and professional books which focus on the theory and applications of Cyber Security, including Data Security, Mobile and Network Security, Cryptography and Digital Forensics. Topics in Prevention and Threat Management are also included in the scope of the book series, as are general business Standards in this domain.

Books published in the series include research monographs, edited volumes, handbooks and text-books. The books provide professionals, researchers, educators, and advanced students in the field with an invaluable insight into the latest research and developments.

Topics covered in the series include-

- Blockchain for secure Transactions
- Cryptography
- Cyber Security
- Data and App Security
- Digital Forensics
- Hardware Security
- IoT Security
- Mobile Security
- Network Security
- Privacy
- Software Security
- Standardization
- Threat Management

For a list of other books in this series, visit www.riverpublishers.com

Artificial Intelligence and Blockchain in Digital Forensics

Editors

P. Karthikeyan
National Chung Cheng University, Taiwan

Hari Mohan Pandey
Bournemouth University, United Kingdom

Velliangiri Sarveshwaran
SRM Institute of Science and Technology, India

LONDON AND NEW YORK

Published 2023 by River Publishers
River Publishers
Alsbjergvej 10, 9260 Gistrup, Denmark
www.riverpublishers.com

Distributed exclusively by Routledge
4 Park Square, Milton Park, Abingdon, Oxon OX14 4RN
605 Third Avenue, New York, NY 10017, USA

Artificial Intelligence and Blockchain in Digital Forensics / P. Karthikeyan, Hari Mohan Pandey and Velliangiri Sarveshwaran.

Routledge is an imprint of the Taylor & Francis Group, an informa business

ISBN 978-87-7022-688-2 (print)
ISBN 978-10-0084-806-9 (online)
ISBN 978-1-003-37467-1 (ebook master)

Contents

Preface

This book can provide a wide-ranging overview of how AI and Blockchain can be used and solve the problem in digital forensics using advanced tools and applications available in the market. This book is a collection of 15 chapters authored by academics and practitioners worldwide. The contributions in this book aim to enrich the information system discipline by providing the latest research and case studies from around the world.

This book describes AI and Blockchain frameworks to make the digital forensic process efficient and straightforward. AI feature helps determine the contents of a picture, detect spam email messages and recognise swatches of hard drives that could contain suspicious files. Blockchain-based lawful evidence management scheme supervises the entire evidence flow and the court date.

All thc books publishcd carlicr by diffcrcnt authors only addrcss digital forensics in general. Hence, it is decided to propose a book which not only discusses digital forensics and also solves those problems with the help of artificial intelligence and blockchain. This book is starting point for the researchers how they can apply AI and blockchain in the digital forensics. So, this is very useful for students, academicians and research scholars to explore further in their field of study. It is very much opted for readers who seek to learn from examples.

Acknowledgment

Editors are also grateful to graduate students and young researchers in the Dept. of Computer Science and Information Engineering (National Chung Cheng University, Taiwan), Department of Computational Intelligences Institute of Science and Technology Kattankulathur Main Campus Chennai, India and Data Science and Artificial Intelligence Department, Bournemouth University,

For help with preparing the manuscript, we are thankful to Junko Nakajima (River Publishers) for her encouragement throughout the project. Our sincere apologies to everyone we might have overlooked.

P. Karthikeyan,
National Chung Cheng University, Taiwan.

Hari Mohan Pandey,
Bournemouth University, United Kingdom

Velliangiri Sarveshwaran,
SRM Institute of Science and Technology, India.

List of Contributors

Anandaraj S. P., *Associate Professor, Department of Computer Science and Engineering, School of Engineering, Presidency University, Bangalore, India*

Angelin Gladys, *Lecturer, Department of Information Technology, University of Technology and Applied Sciences, Ibri, Oman*

Asadi Srinivasulu, *Data Science Research Lab, BluCrest University- 91016, Monrovia, Liberia*

Ashutosh Raj, *Assistant Professor, Dept. of ISE, BMS Institute of Technology and Management, Avalahalli, Yelahanka, Bangalore – 64, India*

Atul Tomar, *Assistant Professor, Dept. of ISE, BMS Institute of Technology and Management, Avalahalli, Yelahanka, Bangalore – 64, India*

Ayush Prakash, *Assistant Professor, Dept. of ISE, BMS Institute of Technology and Management, Avalahalli, Yelahanka, Bangalore – 64, India*

Biswas R., *Applied Optics and Photonics Lab, Department of Physics, Tezpur University, Tezpur-784028, India*

Biswas S., *Department of English, Amguri College, Amguri-785680, India*

Chetan Singh, *Assistant Professor, Dept. of ISE, BMS Institute of Technology and Management, Avalahalli, Yelahanka, Bangalore – 64, India*

Danu R., *Research Scholar, SRM Institute of Science and Technology, Kattankulathur, Chennai 603203, India*

Deepakanmani S., *Department of IT, Sri Krishna College of Engineering and Technology, Coimbatore, India*

Dinesh Kumar R., *Professor, CSE Department, Siddhartha Institute of Technology and Science, Hyderabad, India*

Durga S., *Associate Professor, Department of IT, Sri Krishna College of Engineering and Technology, Coimbatore, India*

Esther Daniel, *Associate Professor, Department of CSE, Karunya Institute of Technology and Sciences, Coimbatore, India*

Fahmina Taranum, *Professor, Department of Computer Science and Engineering, Muffakham Jah College of Engineering and Technology, Banjara Hills, Hyderabad-500034, India*

Getzi Jeba Leelipushpam Paulraj, *Associate Professor, Department of CSE, Karunya Institute of Technology and Sciences, India*

Gurram Venkata Siva Nandan, *Senoir Risk and Reliability Engineer, Phillips-Medisize A/S, Struer, 7600 Denmark*

Immanuel Johnraja Jebadurai, *Professor, Department of CSE, Karunya Institute of Technology and Sciences, India*

Ishi Saxena, *Department of Computing Technology, SRMIST, KTR, India.*

Jebaveerasingh Jebadurai, *Assistant Professor, Department of CSE, Karunya Institute of Technology and Sciences, India*

Karthika B., *Assistant professor, IT Department, PSNA College of Engineering and Technology, Dindigul, India*

Karthikeyan P., *Post-Doctoral Researcher, Dept. of Computer Science and Information Engineering, National Chung Cheng University, Chiayi, Taiwan-62102*

Kavitha N., *Indra Ganesan College of Engineering, Trichy, Tamil Nadu, India*

Kavitha S., *Associate Professor, SRM Institute of Science and Technology, Kattankulathur, Chennai 603203, India*

Kesavamoorthy R., *Associate Professor, Department of Computer Science and Engineering, CMR Institute of Technology, Bengaluru, India*

Mahesh T. R., *Associate Professor, Department of Computer Science and Engineering, Faculty of Engineering and Technology, Jain Deemed-to-be University, Bangalore, India*

Malhotra S., *Shaheed Bhagat Singh Evening College, University of Delhi, India*

Maria Nancy, *Department of Computing Technology, SRMIST, KTR, India*

Mary Neeba T., *Assistant Professor, Department of ECE, Karunya Institute of Technology and Sciences, Coimbatore, India*

Muneshwara M. S., *Assistant Professor, Dept. of ISE, BMS Institute of Technology and Management, Avalahalli, Yelahanka, Bangalore – 64, India*

Murugesh V., *Assistant Professor, College of Informatics, Bule Hora University, Ethiopia*

Niraja K.S., *Assistant Professor, Department of Information Technology, BVRIT Hyderabad College of Engineering for Women, Bachupally, Hyderabad, India*

Pedro C. Flores, *Lecturer, Higher College of Technology, Dubai UAE*

Rajesh Kumar V., *Assistant Professor, Department of Electrical and Electronics Engineering, Sir M Visvesvaraya Institute of Technology, Bengaluru, India*

Ravi Chand S., *Professor, ECE Department, Nalla Narasimha Reddy Group of Institutions, Hyderabad, India*

Ruba Soundar K., *Mepco Schlenk Engineering College, Sivakasi, Tamil Nadu, India.*

Sabapathi V., *Research Scholar, Department of Computer Science and Engineering, SRM Institute of science and technology, Kattankulathur, Chennai-603 203, India*

Salaja Silas, *Professor, Department of CSE, Karunya Institute of Technology and Sciences, India*

Saranya S., *Research Scholar, Department of Computing Technologies, School of Computing, SRM Institute of Science and Technology, India*

SathisKumar T., *Saranathan College of Engineering, Trichy, Tamil Nadu, India*

Sathya K., *Assistant Professor, Department of CT/UG, Kongu Engineering College, Tamilnadu, India*

Selvin Paul Peter J., *Associate professor, Department of Computer Science and Engineering, SRM Institute of science and technology, Kattankulathur, Chennai-603 203, India*

Shanmuga Priya S., *New Horizon College of Engineering, Bengaluru, Karnataka, India*

Shanmugarathinam G., *Associate Professor, Presidency University, Bengaluru, India*

Sheetal, *Assistant Professor, Presidency College, Bengaluru, India*

Swetha M. S., *Assistant Professor, Dept. of ISE, BMS Institute of Technology and Management, Avalahalli, Yelahanka, Bangalore – 64, India*

Usha G., *Associate Professor, Department of Computing Technologies, School of Computing, SRM Institute of Science and Technology, India*

Vani Rajasekar, *Assistant Professor, Department of CSE, Kongu Engineering College, Tamilnadu, India*

Veena S., *Department of Computing Technology, SRMIST, KTR, India*

Velliangiri S., *Assistant Professor, Department of Computational Intelligences Institute of Science and Technology Kattankulathur Main Campus, India*

Vinayakumar Ravi, *Assistant Research Professor in Artificial Intelligence at Prince Mohammad Bin Fahd University, Saudi Arabia*

Vinoda Reddy G., *Professor of Computer Science and Engineering (AI & ML) CMR Technical Campus, Hyderabad, India*

Vinoth N. A. S., *Department of Computing Technology, SRMIST, KTR, India*

List of Figures

List of Tables

List of Abbreviations

AES	Advanced encryption standard
AI	Artificial intelligence
ANN	Artificial neural networks
ANNs	Artificial neural networks
API	Application programming interface
ARP	Address resolution protocol
ATM	Automated teller machines
BC	Blockchain
BoW	Bag of words
CAM	Camera
CC	Chain of custody
CCMS	Component content management system
CMS	Content management system
CNN	Convolutional neural network
CNNIC	China internet network information centre
CoC	Chain of custody
CSEM	Child sexual exploitation material
DAM	Digital asset management system
DBMS	Data base management system
DCMS	Digital content management system
DCMS	Document content management system
DF	Digital forensics
DFIR	Digital forensics incident response
DL	Deep learning
DLT	Distributed ledger technology
DNA	Deoxy-ribonucleic Acid
DOS	Denial of service
DW	Digital witness
ECC	Elliptic Curve cryptography
ECMS	Enterprise content management system

eDiscovery	electronic discovery
ESI	Electronic stored information
ESN	Echo state networks
ETH	Ether
EXIF	Exchangeable image file format
FDA	Forensic data analysis
FWHM	Four way handshake methodology
GDPR	General data protection regulation
GPS	Global positioning system
GWI	Global system web index
HTML	Hyper text markup language
IA	Internet addiction
ICO	Initial currency offering
IOC	Indicators of compromise
IoFT	Internet of forensic things
IoT	Internet-of-things
IP	Internet protocol
IPFS	Inter planetary file system
IR	Incident response
IT	Information technology
JPEG	Joint photographic experts group
KAPE	Kroll artifact parser and extractor
KAS	Key authority service
LSTM	Long short-term memory
MAC	Medium access control
ML	Machine learning
MLP	Multilayer perceptron
NLP	Natural language processing
NSFW	Not safe for work
OCG	Organized criminal groups
OGA	Online gaming addiction
OPI	Online predator identification
OSN	Online social network
PC	Personal computer
PDA	Personal digital assistant
PDF	Portable document format
PoW	Power of work

RAM	Random access memory
RBAC	Role based access control
RDF	Resource description framework
RNN	Recurrent neural networks
RSA	Rivest–Shamir–Adleman
SDN	Software defined networks
SHA	Secure hashing algorithm
SMPs	Social media platforms
SNMP	Simple network management protocol
SNS	Simple notification service
SVM	Support vector machine
TCP	Transmission control protocol
UDP	User datagram protocol
UPA	User permission agreement
URL	Uniform resource locator
USB	Universal serial bus
VPKI	Vehicular public-key infrastructure
VSM	Vector space model
WCMS	Web content management system
Wi-Fi	Wireless fidelity
XAI	Explainable artificial intelligence
XML	Extensible markup language

1

Digital Forensics Meets AI: A Game-changer for the 4th Industrial Revolution

S. Malhotra

Shaheed Bhagat Singh Evening College, University of Delhi, India
Email: suzaneemalhotra@commerce.du.ac.in

Abstract

Digital forensics is defined as the actual process of recognizing, mining, documenting, and safeguarding proofs, testimonials or evidence of a digital nature, which are admissible in a court of law, and play a massive role in maintaining the rule of law. The present IT (information technology) and AI (artificial intelligence) have made digital forensics a part of mainstream business, whose popularity is growing with each passing day. The growing relevance of digital forensics calls for a sound understanding of this concept. Also, the evolving nature of AI calls for a special mention, understanding of the role that AI plays in the domain of digital forensics and how the latest AI trends are impacting on digital forensics. This chapter will use all this pertinent information to reach the goal post of providing a sound conceptual understanding of digital forensics and AI allied techniques and trends. The journey towards that goal post will be an insightful and enriching one.

1.1 Introduction

We are currently witnessing some unprecedented and emerging technologies that are impacting our lives, and will impact generations to come, as part of developments in the fourth industrial revolution [42]. The wonders of technology keep coming up in new and unanticipated ways. One such development is the technology of artificial intelligence or AI, which has brought in great changes in human history. AI refers to the capability of machines to learn and adapt by becoming better at making predictions for human behaviour based on the data consistently fed into their systems. Today's whole business world and many of its allied activities revolve around the cognitive decision-making ability and power of AI-based datasets [28, 42, 43].

In this new era of digitization, AI in recent years has made a spot for itself in our lives because AI has made such massive progress, driven by the exponential rise in computing power, impressive growth of big data and breakthrough landmarks in the field of algorithms [42]. One such field where the usage of AI on datasets is evolving is digital forensics [4]. This chapter aims to present an overview of the discussion regarding the two concepts of digital forensics and AI, as well as indicating the signposts of the journey ahead; from where we are at the current time to where we will be taken along this ever-evolving journey.

1.2 Digital Forensics

"Peek-a-boo, I'm watching you", "your footprints are being trailed", "you are under surveillance", and many more such phrases are a fact of the digital age of our times. We all are under a radar, a digital radar, where the binary language of 1s and 0s is scrutinizing, scanning, and staring at us through the means of our smartphones, call and messaging history, financial transactions, emails and also the innocent web searches we often make to look for information [20]. Everything leaves a digital trail for our privacy to be exposed via our massive dependence on the latest tools of technology and development; "we are drowning in electronically stored information" [20].

The era of social media has changed the functioning of courts, from traditional paper-based mechanisms to revolutionary digital evidence, which offer opportunities and challenges simultaneously [3, 20]. The growing dependence on the technological means and tools have brought in a tide of change in the domains of law and the legal system, where each day piles of digital evidence not only find their way in and out of courts but also play a critical role in assuring justice and the "rule of law" [20]. Digital evidence differs from traditional paper-based evidence, which calls for an inevitable transformation in the procedures and processes of our legal system mechanisms as well, be it related to administrative, civil or criminal nature, for digital evidence to be admissible in the court of law: "the best scientific evidence in the world is worthless if it is inadmissible in a court of law" [20].

Forensic science refers to the "application of science to solve a legal problem" [20]. Forensic science is a perfect blend of the domains of law and science. It relies on both pillars to be understood and applied [20, 39].

While forensic science has been around us, guiding and helping at length for decades now, digital forensics as a concept is at an infant stage trying to earn its share of reputation among other forensics disciplines (like DNA or toxicology) since as a sub-field of forensics its standards, principles and practices are being developed [20].

Ken Zatyko, in Forensic Magazine, describes digital forensics as "the application of computer science and investigative procedures for a legal purpose involving the analysis of digital evidence after proper search authority, chain of custody, validation with mathematics, use of validated tools, repeatability, reporting, and possible expert presentation" [44]. Årnes has defined digital forensics as "the use of scientifically derived and proven methods toward the preservation, collection, validation, identification, analysis, interpretation, documentation, and presentation of digital evidence derived from digital sources to facilitate or further the construction of events found to be criminal, or helping to anticipate unauthorized actions shown to be disruptive to planned operations" [19].

In simple terms, digital forensics is the application of forensic science to any format or file of digital information for digital investigation [19]. It is a broader concept, not restricted to merely simple computers, laptops, mobile phones, and other network and cloud systems and architecture, but also includes analysis of computer files in image, video or audio formats [20, 35]. It is carried out with the intent of "authenticity, comparison, and enhancement" [20].

Thus, the art and science that aims to abstract evidence of a digital nature out of digital, technological or electronic gadgets and devices like mobile phones, tablets, computers, laptops, server and related network equipment or other periphery devices like pen-drives, hard disks, etc., is precisely termed as digital forensics [5]. Digital forensics lends its hand to forensic science team members, making it possible for them to apply advanced and sophisticated techniques and tools to solve and resolve complex cases related to digital issues and problems [9].

One cannot gain a comprehensive understanding of digital forensics without scratching one's head for some binary 1s and 0s that serve as the foundation for digital forensics and a thorough and intricate understanding of the functioning of computers [20]. Two sister concepts often referred to in digital forensics are digital archaeology and digital geology [19]. While the former indicates the trails and traces of digital footprints owing to the human's actions of exploring computer systems, the latter refers to the trails and traces of digital footprints that occurred and were created by the default actions of computer processes themselves [45].

Establishing objective and factual answers to legal problems is the primary function of digital forensics or practitioners working in this area. Within a structured inquiry, a proper track of the "chain of evidence", that is, what happened, how it happened, when it happened, and how does the digital device or gadget relate to this chain and who was the preparator, are a few questions asked by digital forensics [17].

Digital forensics means gathering facts and data, followed by a systematic analysis of such data by investigators to unravel the what, how and who of the changes made [11]. Digital forensics is not always used regarding criminal activity, sometimes digital forensics is put to use for important purposes, such as recovering data from crashed or corrupted devices, or formatted operating systems [31].

Recovery of critical data in compliance with the legalities, rendering such data admissible in legal proceedings, refers to digital forensics, dubbed computer forensics or cyber forensics [12, 34]. The scrutiny in digital forensics provides crystal clear answers to what data or device was tampered with, the login and log out records, ways and means of tampering or what malware is at play [13].

1.2.1 Growing need for digital forensics

With the help of digital forensics, the integrity of the rule of law in our civilized societies can be maintained [17]. In particular, in the ever-changing, ever-rising and ever-evolving digital age of our times, the importance and relevance of digital forensics have become more important than ever, offering us a mechanism for safeguarding ourselves from the grip of rising cyber-crime and attacks of various types [17, 36].

The average person remains unaware of what information they keep adding to digital media and devices in ordinary life. For example, in the event of an accident, there is a lack of substantial evidence per se. In such an event, one can use the digital computer system installed in the cars of the parties concerned to probe into the driving history and patterns of driving behaviour, yielding some critical digital evidence which can help in legal issue resolution or establishment of justice.

From simply fighting against vicious online or cyber-crimes to protection from "denial of service" attacks, the need for digital forensics is immense [5, 18, 19]. Not only do digital forensics and evidence enable tackling digital or cybercrimes, but ordinary and petty criminal activities like burglary or complex issues like assault and murder [18]. Application of multi-level data governance and layers of network security by business organizations can be

Figure 1.1 Digital forensics process.

helpful the streamlining of digital forensics processes when necessary [12]. From basic browsing history to email logs, all offer great assistance to investigators to track cybercriminals and attackers' digital trails and footprints [17].

1.2.2 Process of digital forensics

The digital forensics process comprises standard procedures followed in a systematic approach to filter and sort the data in a point-wise fashion. The entire process depends on the background of the investigation, the media or devices under the investigation scanner or the information sought by the investigators [6]. Generally, the steps entailed in the digital forensics process are depicted in Figure 1.1 [6].

Let us discuss these steps in detail.

Step 1: Identification

The identification step in the process of digital forensics deals with spotting evidence or files and assessing them as to what, where, when, by whom, why and how such evidence or files were stored in different electronic storage media like computers, laptops, mobile, hard disks, etc. [6]. A forensic image or a digital copy of any digital file or evidence is often made for further analysis and study while keeping the original records in a secure and pristine form isolated from others. Sometimes, publicly available data like posts on Facebook, Instagram, Twitter, etc. can also serve as potential evidence for scrutiny.

Step 2: Preservation

To preserve means to isolate and secure any preciously important subject matter. In digital forensics, preservation involves isolating, securing and preserving data files that are very important for the investigation [6]. By securely isolating the data, this process aims to prevent any third party from accessing such data or devices, disregarding the possibility of any data tampering.

Step 3: Analysis

Under the analysis step, the pieces of digital evidence or data are put under intensive study by the various digital forensic scientists to draw meaning or conclusions [6]. For such an intensive study, a number of iterations or systematic rounds of examination of the data are needed to lend sufficient justification to the interpretations drawn out of the data [1, 12]. A few AI tools like "Basis Technology's Autopsy" (examining hard drives), "Wireshark" (assessing the network protocols), "mouse jiggler" (to examine the volatile computer memory) are useful in analysing the digital evidence.

Step 4: Documentation

It is also essential to track all the data and evidence gathered in an investigation [6]. Such tracking and data maintenance can play a significant role in the re-creation and review of the crime scene at later stages. Thus, the documentation step takes care of this necessity by keeping a detailed record of all the data files and documents related to the crime scene, photographs, sketches and crime-scene mapping.

Step 5: Presentation

The presentation step summarises the entire findings and conclusions drawn from the digital files and evidence [6]. Such a summarised form of findings can be used in the legal proceedings or the court of law for the judges or jury's reference help, helping them arrive at appropriate verdicts for the lawsuits. However, sometimes the summarised findings are also presented in a very simplified manner using layman's abstracted terminologies if the understanding of such findings is necessary for the general public.

1.2.3 Advantages offered and limitations confronted by digital forensics

Digital forensics as a domain of study has much to offer, including many advantages. However, this domain does not entirely come free of limitations. A few of the advantages offered and limitations conferred by this domain of science are discussed in Table 1.1.

1.3 AI and Digital Forensics

1.3.1 Contribution of AI in the realm of digital forensics

The field of digital forensics requires intensive number-crunching and computational aspects, instantly and at the same time intelligently. With artificial intelligence on our side, many complex problems can be readily solved or handled in real-time [28]. Some essential requirements like a storage facility for large datasets, unison of mechanisms and the technical sophistication of the

Table 1.1 Digital forensics advantages and limitations.

Advantages offered	Limitations conferred
1. Digital forensics help to maintain the integrity of a computer system and digital architecture [23].	1. The cost of procuring and preserving digital evidence is high. Thus, it is costly for small and medium firms [23].
2. Digital forensics helps to penalise wrong-doers in a court of law by providing substantial evidence [15].	2. Data leaks or breaches may happen when cloud storage is used to back-up data [23].
3. Any kind of comprise in the computer or network security can be caught using digital forensics [17].	3. Any tool or method used in digital forensics, if deemed unfit in the eyes of the law, is disregarded and rejected [20].
4. This process helps to draw out essential meanings and inferences from the digital pieces of information [9].	4. Suppose the competent or investigating officer does not have the technical or desirable know-how about some facts. In that case, it might not yield the desired output [8].

methods can all be efficiently addressed on a real-time basis the moment AI is introduced into the game of digital forensics [2, 28]. Thus, the use of AI in handling the issues involved in digital forensics is the call of changing times.

1.3.1.1 Knowledge representation

AI is empowered with a convenient knowledge representation methodology, which means a way to represent or depict the knowledge or information of interest among pools of data [7, 28]. Suppose the knowledge representation methodology is coupled with ontology. In that case, one can also trace the reasons behind such depiction or presentation to make intelligent sense of the data [7, 28]. Two popular methodologies that do this are Extensible Markup Language (XML) and Resource Description Framework (RDF), which can be successfully put to use to make the digital forensics data standardised [13, 28].

Making the data-sets standardised for digital forensics yields essential pieces of information from the different data sources and automatically prepares the information in a proper manner amenable to be used for discussion or reference. Such standardised data collated together using AI techniques would become a reusable data repository for the future, becoming an essential asset, handling the issues often faced in the storage and retrieval stage.

1.3.1.2 Reasoning process

AI can also lend its support in the process of reasoning that makes use of algorithms on datasets to make decisions from it [7]. AI techniques are branched into symbolic and sub-symbolic types based on character processing for typical reasoning processes [28]. While the former requires a pre-existing expert system resulting in quicker decision outputs [28]. The latter handles any problem with a fresh approach, treating each act of data handling as unique and different [28]. Although the symbolic type presents decisions quickly and efficiently, there can be instances of systematic errors because the base rules are disregarded and the the rule of thumb is used. Thus, the non-symbolic type is used to avoid errors during analysis and to address the problems with the objectives framed. However, for minor or low-level issues, it becomes unsuitable to use because it is a time-consuming process.

1.3.1.3 Pattern recognition

Spotting clusters of similar data in a large pool of data is accomplished by the pattern recognition technique of AI [7, 21]. Under this technique, a search for pattern match or similarity is made across all data, matching the data with a high probability of being close, and posting the matched clusters that are classified [21, 28]. AI neural nets and decision trees help a great deal in accomplishing this task. They can help identify the pattern sets in a massive amount of data for tracing the trail of cyber attackers or suspicious malware [28].

1.3.1.4 Knowledge discovery

Data mining is a blend of different probabilistic and statistical methods that can play an essential role in adding to the understandability of the data [28]. Data mining explores the data to discover the various relationships among the pieces of information in datasets. This technique can play a significant role in the primary assessment for digital forensics, helping in pattern recognition [38]. The data mining technique of AI can also prove to be very helpful in digital forensics [7, 26].

1.3.1.5 Adaptation

Machine learning can be used efficiently for adaptation, that is, for data refining and learning of data, especially of any damaged or corrupted data source [28]. Thus, managing these repairs to explore the amenable or insightful data can help digital forensic experts.

1.3.2 Different variants of AI-based digital forensics

Supplementary aid lent to the digital forensics investigation by artificial intelligence technology at play in the form of different variants of digital forensics has a crucial role in verifying and validating the results and findings to maintain the system integrity [40]. Different computer forensic examinations are used for varied and specific aspects of information and evidence collection using AI tools and mechanisms [33]. Some of the critical variants of AI-enabled digital forensics are as follows:

1. Disk forensics: This involves a search, extraction and re-fragmentation of data sources from digital devices by looking for deleted, active, modified, or corrupted files [41].
2. Network forensics aims to monitor and examine the different types of data and network traffic to track important sources and information as legal evidence [7, 41]. The application of a firewall or intrusion detection systems is done for this purpose.
3. Wireless forensics: Similar to network forensics, wireless forensics aims to offer the tool methods required to scan and study the data and network traffic related to wireless devices [41].
4. Database forensics: The extensive examination and study of different databases and meta-data come under database forensics [7, 41].
5. Malware forensics: To identify the potential source of malicious codes, viruses, worms (for example, programs like Trojan horses ransomware, among other viruses) and to study their features to offer future protection is the role of malware forensics [41].
6. Email forensics: Recovery and study of important information as found in email logs or related mailing lists or calendar schedules is the aim of email forensics [7, 41].
7. Memory forensics: Memory forensics aims to collect data from a computer system, cache memory, the RAM in raw form and then draw out understandable information from it [41].
8. Mobile phone forensics: Detailed scanning and examination of mobile phones to retrieve the phone data, SIM details and contacts, the call and messaging records and other audio-video files [41].

1.3.3 AI techniques used by digital forensics investigators

AI seems to provide digital evidence investigators with diverse techniques and proprietary applications to examine the critical sources of information they have. From searching for hidden files or folders to assessing the unallocated disk storage in search of any copy of deliberately deleted or encrypted files, the computer forensic investigators apply various AI techniques, their know-how and expert knowledge in their quest for answers. Some of the popularly used standard AI techniques used in digital forensics are as follows:

1. Reverse steganography: Steganography is made to conceal data in any digital file format. Thus, the use of reverse steganography is made by computer forensic experts using data hashing functions to identify the potential information that has been altered or secretly hidden [40].

2. Stochastic forensics: Many times, when dealing with digital pieces of information, the use of digital artifacts is made to uncover some clues from the data. However, the use of such artifacts changes some essential data attributes. Thus, in stochastic forensics, the data experts aim to assess and recover the information without using artifacts of any type [40]. This technique is most widely used in data breach occurrences when the suspect is usually an insider.

3. Cross-drive analysis: This technique uses several different computer systems and devices to collate and complement the findings to uncover and assess the critical data [40]. Suspicious events or information are compared across with other to yield some context from them. It is also known as anomaly detection.

4. Live analysis: This technique makes an assessment of files or data inside of a system on a real-time basis, when the machine is on its run-time, then examination is made of the files inside it [40]. The results of such a technique are highly volatile. They are generally stored in the RAM or the system's cache memory.

5. Deleted file recovery: Under this technique, a search is made across the scrapes, memory fragments or recycle bins of the computer for any file removed from one location but left its trace elsewhere [40]. It is similar to a scavenging hunt, also known as file or data carving.

Although some of the AI techniques can be at the beginning level, implementing these techniques in digital forensics is the imminent need of the hour [16]. From automating the forensic analysis process [27] to adding an intelligent expert opinion, AI can also play a positive role in rendering justice and maintaining law and order in the digital domain [28].

1.3.4 Deep learning tools and techniques helping in the domain of digital forensics

Deep learning techniques have evolved as a new set of defences against fake news or spam on the growing online social networks, strengthening the hold of digital forensics on digital information [49]. Deep learning has become a great tool in the hands of AI to deal with the budding issue of misinformation detection, especially in the booming age of social media apps [49]. There is wide diversity in the deep learning techniques available to probe the misinformation detection, including the likes of discriminative, generative and hybrid models, which further sub-diversify into different tranches of models like convolutional neural networks, recurrent neural networks, recursive neural networks, deep belief networks, variational auto-encoders, etc. [49].

The use of deep learning and supervised machine learning also empowers digital forensics to detect and diagnose the post-processing of digital videos with different formats that may have been manipulated [50]. Such empowering techniques yield action after getting the data set ready for extraction and supervised machine learning model preparation [50].

The deep learning know-how not only helps in preventing misinformation or post-processing video analysis, but also plays a role in preventing the rising cases of depression or anxiety symptoms among people plagued with growing internet addiction, with its techniques aiding in labelling the different human emotions identified as happy or gloomy [51]. In such a scenario, deep learning indeed represents that AI comes to the aid of humans.

1.4 Latest AI Trends Impacting Digital Forensics

Digital forensics is an up and coming field of study that combines the law, computer science and technology disciplines [20, 25]. The core of artificial intelligence resides in machines mimicking human-like intelligence via a programmed set of instructions to think and react like humans while performing tasks [14, 22]. Today's contemporary world has witnessed the well-established benefits of data-driven resources and technologies associated with them. The unrelenting pathway of technological progress has modified the existing boundaries or limits to things, and has set new ones, giving birth to changed realities, behaviours and trends [43]. This data-driven pathway of technological development has led to the rise of new evolutionary trends into the forays of AI-enabled digital forensics [37], which are discussed at length in this section.

1.4.1 AI has taken a leap from novelty to necessity

Although AI is not new and has been around for some time, the reason behind its sparse usage in the past is owed to the fact that previously the infrastructure, intellectual property, technology and skills allied to AI were expensive and could only be used by only a few influential organizations and institutions [43].

However, with the emerging cloud storage mechanics [29], where the immense pool of data is available in a ready-made manner, there has been a surge in the applicability of AI as a mainstream necessity over novelty [23]. With the utility to pay for as much data content as required, the budget restrictions which earlier put the smaller and medium organizations out of the AI race are now done away with. This has been aptly termed as a form of "plug-and-play application" in [43], referring to the ease of the deployment of AI data-driven solutions to any organization or for any purpose. AI is no longer a mere mechanism of data entry or simple automatic tasks that are repetitive. It has the potential to make intelligent, more innovative and strategic decisions based on data feed.

1.4.2 Data-driven AI can generate valuable content

AI and its allied techniques can create art, music scores and computer programs [43]. With more refined, sophisticated machine learning algorithms being developed, AI has proved to be worth every penny that has been devoted to the continuous machine runs for purposes like making analytical reports for products or services under review, or choosing graphics and designs for video editing, or preparing a report for the digital health of a computer system or network. Not only are all these tasks accomplished at lower cost but also in an efficient and targeted manner.

1.4.3 Smaller datasets are as amenable as big data

The traditional belief about AI and data-driven techniques is that they are considered to be only amenable and applicable to big data sets collected from a humungous pool of information. However, the revolutionary and evolving AI technology has proved that the technologies also need to adapt swiftly in times of the changed realities of our times. In the pandemic of 2020, when the world was facing something new and with very little data available, AI indeed indicated that in times of need could be run on a small set of data successfully and informatively when that is all that is available [43].

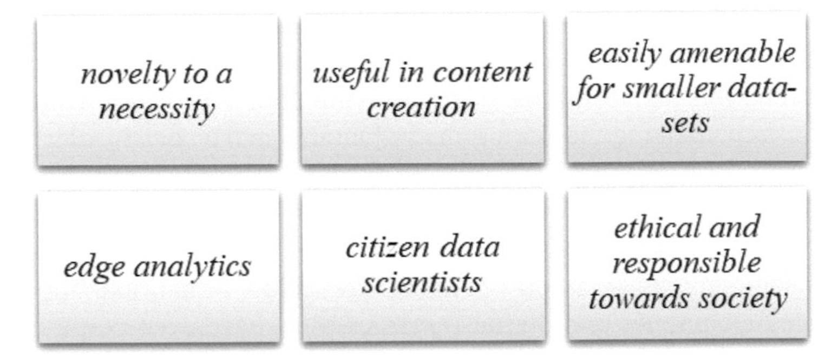

Figure 1.2 Latest AI trends impacting digital forensics.

1.4.4 Edge analytics: An upcoming AI trend

Edge computing or edge analytics have become buzzwords recently [43]. Under this technique, the intensive computational process is done at a point closer to the data pool source or even in the same place that the data is stored or collected. Thus, unlike cloud computing, which fetches its data from far-off data repositories, edge analytics focuses on real-time AI applications like self-driven cars or robot nurses. Under edge analytics, the speed of computing and decision delivery is the essence to making actions seamless and quick, just like a rash human judgement would be in times of crisis or shock [43].

1.4.5 Citizen data scientists: The next big thing under AI

A recent trend in AI is the concept of citizen data scientists which seeks to overcome the paucity of professionally trained people across the domains where it can be successfully put to use. As described by Gartner, a citizen data scientist may not be academically or professionally trained in the field of AI or data-driven technologies but has enough know-how to come up with plausible solutions for regular and routinised tasks [46]. Thus, any person who possess limited or essential skills for data-modelling or computing, can take decisions based on that knowledge and one who is constantly learning to improve his/her skills, would be considered a citizen data scientist [43]. This particular trend can help digital forensics where, currently, AI skills are not as refined as they ought to be for rendering help.

1.4.6 AI has an ethical and responsible role in society

With the domain and dimensions of AI being so large, AI must render an ethical and responsible role to society. In addition, due to the susceptibility of AI technology being used by wrong-doers or criminals, AI must deliver its ethical role for the greater good of society. Moreover, what could be a better way of doing this than using the power of AI in digital forensics. It would not only help in penalizing the wrong-doers but also make social order just and equitable.

AI-enabled digital forensics have the potential to save individuals, businesses and even governments from potential losses involving not only money but invaluable intangible assets or strategic resources of national importance. AI-enabled techniques and digital forensics tools have even come in handy to protect a considerable number of Apple trade secrets from being stolen and exploited [40]. A Chinese engineer working at one of Apple's autonomous car divisions gave notice of his voluntary retirement from the company stating his intention to move back to his homeland and work in an EV manufacturing firm. However, his immediate supervisor was suspicious of his intentions. When the company's security team traced his digital trail across the company's systems and networks, it was discovered that he had illicitly downloaded the company's trade secrets worth millions from the confidential company databases that were available to him. Thus, he was rightly indicted by the FBI in 2018, and all this was due to AI-enabled digital forensics tools and methods.

1.5 Challenges and the Road Ahead

1.5.1 Key challenges to be addressed

Despite the field of digital forensics developing and becoming the requirement for the hour, there remain quite a few challenges ahead [8], which are discussed as follows.

1.5.1.1 Heterogeneity, resulting in lack of standardization

With the continuous evolution of digital devices, heterogeneous kinds of hardware and software platforms and no designated standards or procedures dealing with sources and investigation raises the question on the field being built-up using fragmented pieces from here and there with no prescribed set of standards, which generally describes any field of science [8].

1.5.1.2 AI can be a double-edged sword

AI and its allied techniques can function as a double-edged sword. On one hand, when used to catch criminals, it can cut for us. However, when the same techniques are in the hands of criminals, it can also cut against us.

The potential use of AI techniques against us is popularly known as 'anti-forensics' [47].

1.5.1.3 Privacy-preserving and legitimacy outcry

Going digital has made the thin line between anyone's personal and private lives gradually disappear. Loads of data is collected across various social networking sites. However, the real question is whether gaining access to such data for tracing an individual indeed hinders their privacy rights or not [47]. The privacy concerns exist, but the growth of the dark web has raised questions about the legitimacy of the source of the data as procured under the part of digital forensics investigation [47]. If any data is gained from such dark-web platforms, it would be challenging to execute the investigation legally and may become subject to law violations.

1.5.2 Road to the future

Digital forensics need to be strengthened enough to "reverse the fate" of any cyber-crime or tampering and explore its share of opportunities to fight the challenges ahead and become a robust science for generations to come [8].

A code of procedural handling of investigations reported under the digital domain is a must [8, 47]. Suppose the higher authorities diligently take this step. In that case, it will not only help ward off the challenge of heterogeneity in platforms and devices, but also the mechanics of how to carry out any investigation would be established, making the field of digital forensics similar to an organised body of science.

To fight off the menace of anti-forensics, it would become imperative to use AI and its allied techniques and digital forensics power to end, or at least minimise, the misuse of technology [47]. Also, a buffer mechanism could be adopted that traps the criminal perpetrator in their folly. If the technology or the firewall system can be strengthened enough to make it nearly impossible for the perpetrator to delete or remove the evidential files from the system or network forever or from everywhere, then a great deal of anti-forensics activity can be kept in check.

Yes, the privacy and legitimacy debates are a matter of civil rights, but so is digital forensics. Digital forensics intends to fight the digital criminals who are the reason behind the discomfort or financial losses of everyday citizens. Thus, if the situation demands, even the private data of the offenders or suspicious people on social media platforms must be shared, with the principle of the greater good of the maximum number of people in mind. India's new IT rules are also a step in this direction. Moreover, as far as the dark web is concerned, the power of digital forensics must be used to launch a digital criminal investigation against people responsible for creating, maintaining or sharing information on such platforms if even a single clue or tip-off is noticed.

Digital forensics investigations can successfully exploit the new and evolving software-defined networking techniques to provide a buffer abstraction layer during analysis for any suspicious activity trail to confuse the perpetrator in using any leeway for fleeing. Digital forensics can potentially leverage many of the latest wonders of the Internet-of-Things, or IoTs, to make the interaction between the digital and real world seamless [24], for example, usage of nodes to trace the trail of suspicious persons in the form of intelligent devices or sensor objects [8, 48].

1.6 Conclusion

Digital forensics has become a fundamental need for growing digital or cyber extensions in the modern-day world. The growing digital extension has made the modern digital society vulnerable to many digital criminal activities and financial frauds because of the developing technologies in this domain, including the likes of online banking, crypto-currency and blockchain, causing economic loss and mental hazard to the affected parties. Therefore, it has become an absolute need for society to embrace the upcoming wave of digital forensics. Moreover, with the power and competence of AI, the tools and techniques of digital forensics must be engineered to fight back against the cybercriminals and overcome the challenges and difficulties ahead. Intensive investment in the technology, training and tools of digital forensics and AI is imperative for the future success of this upcoming trend. AI with the potential to transform various aspects of the business world can become a total game changer for the 4th industrial revolution if it is combined with the power of digital forensics.

References

[1] Reith M, Carr C, Gunsch G. An examination of digital forensic models. International Journal of Digital Evidence 2002; 1(3): 1–12.

[2] Vasiliev AA, Pechatnova YV, Mamychev AY. Digital ecology: Artificial intelligence impact on the legal and environmental sphere. Ukrainian Journal of Ecology 2020; 10(5): 150–154. DOI: 10.15421/2020_222.

[3] Jang YJ, Kwak J. Digital forensics investigation methodology applicable for social network services. Multimedia Tools and Applications 2015; 74(14): 5029–5040. DOI: 10.1007/s11042-014-2061-8.

[4] Richard III GG, Roussev V. Next-generation digital forensics. Communications of the ACM 2006; 49(2): 76–80. DOI: 10.1145/1113034.1113074.

[5] Garfinkel SL. Digital forensics research: The next 10 years. Digital Investigation 2010; 7: S64-S73. DOI: 10.1016/j.diin.2010.05.009

[6] Flaglien AO. The digital forensics process. Digital Forensics 2017; 13–49. DOI: 10.1002/9781119262442.ch2.

[7] Irons A, Lallie HS. Digital forensics to intelligent forensics. Future Internet 2014; 6(3): 584–596. DOI: 10.3390/fi6030584.

[8] Caviglione L, Wendzel S, Mazurczyk W. The future of digital forensics: Challenges and the road ahead. IEEE Security & Privacy 2017; 15(6): 12–17. DOI: 10.1109/MSP.2017.4251117.

[9] Watson S, Dehghantanha A. Digital forensics: the missing piece of the internet of things promise. Computer Fraud & Security 2016; 2016(6): 5–8. DOI: 10.1016/S1361-3723(15)30045-2.

[10] van Baar RB, van Beek HM, Van Eijk EJ. Digital Forensics as a Service: A game changer. Digital Investigation 2014; 11: S54-S62. DOI: 10.1016/j.diin.2014.03.007.

[11] Mocas S. Building theoretical underpinnings for digital forensics research. Digital Investigation 2004; 1(1): 61–68. DOI: 10.1016/j.diin.2003.12.004

[12] Carrier B. Defining digital forensic examination and analysis tools using abstraction layers. International Journal of digital evidence 2003; 1(4): 1–12.

[13] Garfinkel S. Digital forensics XML and the DFXML toolset. Digital Investigation 2012; 8(3–4): 161–174. DOI: 10.1016/j.diin.2011.11.002

[14] Bhatt P, Rughani PH. Machine learning forensics: A new branch of digital forensics. International Journal of Advanced Research in Computer Science 2017; 8(8): 217–222. DOI: 10.26483/ijarcs.v8i8.4613.

[15] Grigaliunas S, Toldinas J, Venckauskas A, Morkevicius N, Damasevicius R. Digital evidence object model for situation awareness and decision making in digital forensics investigation. IEEE Intelligent Systems 2020; 36(5): 39 - 48. DOI: 10.1109/MIS.2020.3020008.

[16] Mitchell F. The use of Artificial Intelligence in digital forensics: An introduction. Digital Evidence and Electronic Signature Law Review 2010; 7: 35–41. DOI: 10.14296/deeslr.v7i0.1922.

[17] Casey E. Handbook of digital forensics and investigation. 2009. Elsevier. Burlington, MA (USA).

[18] Sachowski J. Digital Forensics and Investigations: People, Processes, and Technologies to Defend the Enterprise. 2018. CRC Press. Boca Raton. DOI: 10.4324/9781315194820.

[19] Årnes A., ed. Digital forensics. 2018. John Wiley & Sons. Hoboken, NJ.

[20] Sammons J. Digital forensics: threatscape and best practices. 2015. Syngress. Waltham, MA.

[21] Mitchell FR. (2014). "An Overview of Artificial Intelligence Based Pattern Matching in a Security and Digital Forensic Context", in Cyberpatterns, ed. C. Blackwell C, H. Zhu (Springer, Cham). DOI: 10.1007/978-3-319-04447-7_17.

[22] Gupta A, Gupta R, Sankaran A. (2021). "Machine Learning Forensics: A New Branch of Digital Forensics" in Confluence of AI, Machine, and Deep Learning in Cyber Forensics, ed. A. Gupta, R. Gupta, A. Sankaran (IGI Global), 47–66. DOI: 10.4018/978-1-7998-4900-1.ch003.

[23] Abdalla PA, Varol A. (2019, June). "Advantages to Disadvantages of Cloud Computing for Small-Sized Business" in 7th International Symposium on Digital Forensics and Security (ISDFS) (IEEE), 1–6. DOI: 10.1109/ISDFS.2019.8757549.

[24] Kruger JL, Venter H (2019, September). "Requirements for IoT Forensics" in 2019 Conference on Next Generation Computing Applications (NextComp) (IEEE), 1–7. DOI: 10.1109/NEXTCOMP.2019.8883615.

[25] Grobler CP, Louwrens CP, von Solms SH. (2010, February). "A multi-component view of digital forensics" in 2010 International Conference on Availability, Reliability and Security (IEEE), 647–652. DOI: 10.1109/ARES.2010.61.

[26] Hoelz BW, Ralha CG, Geeverghese R. (2009, March). "Artificial intelligence applied to computer forensics" in Proceedings of the 2009 ACM symposium on Applied Computing (SAC), 883–888. DOI: 10.1145/1529282.1529471.

[27] Asquith A, Horsman G. Let the robots do it!–Taking a look at Robotic Process Automation and its potential application in digital forensics. Forensic Science International: Reports 2019; 1: 100007. DOI: 10.1016/j.fsir.2019.100007.

[28] Jadhav H. "Artificial Intelligence in Digital Forensics" 2021. Available at: https://community.nasscom.in/communities/emerging-tech/ai/artificial-intelligence-in-digital-forensics.html [accessed September 24, 2021].

[29] Lillard TV, Garrison CP, Schiller CA, Steele J, Murray J. Digital forensics for network, internet, and cloud computing. 2010. Syngress. Burlington, MA (USA).

[30] Li S, Qin T, Min G. Blockchain-based digital forensics investigation framework in the Internet of Things and social systems. IEEE Transactions on Computational Social Systems 2019; 6(6): 1433–1441. DOI: 10.1109/TCSS.2019.2927431.

[31] Kaur R, Kaur A. Digital forensics. International Journal of Computer Applications 2012; 50(5): 5–9.

[32] Ryu JH, Sharma PK, Jo JH, Park JH. A blockchain-based decentralized efficient investigation framework for IoT digital forensics. The Journal of Supercomputing 2019; 75(8): 4372–4387. DOI: 10.1007/s11227-019-02779-9.

[33] Kumari N, Mohapatra AK. (2016, March). "An insight into digital forensics branches and tools" in 2016 International Conference on Computational Techniques in Information and Communication Technologies (ICCTICT) (IEEE), 243–250. DOI: 10.1109/ICCTICT.2016.7514586.

[34] Wiles J, Reyes A. The best damn cybercrime and digital forensics book period. 2011. Syngress. Burlington, MA (USA).

[35] Amato F, Castiglione A, Cozzolino G, Narducci F. A semantic-based methodology for digital forensics analysis. Journal of Parallel and Distributed Computing 2020; 138: 172–177. DOI: 10.1016/j.jpdc.2019.12.017.

[36] Al Mutawa N, Bryce J, Franqueira VN, Marrington A, Read JC. Behavioural digital forensics model: embedding behavioural evidence analysis into the investigation of digital crimes. Digital Investigation 2019; 28: 70–82. DOI: 10.1016/j.diin.2018.12.003.

[37] Wu T, Breitinger F, O'Shaughnessy S. Digital forensic tools: Recent advances and enhancing the status quo. Forensic Science International: Digital Investigation 2020; 34: 300999. DOI: 10.1016/j.fsidi.2020.300999.

[38] Tallón-Ballesteros AJ, Riquelme JC. (2014). "Data Mining Methods Applied to a Digital Forensics Task for Supervised Machine Learning" in Computational Intelligence in Digital Forensics: Forensic Investigation and Applications. Studies in Computational Intelligence, eds. (Springer, Cham), 555: 413–428. Doi: 10.1007/978-3-319-05885-6_17.

[39] Lang A, Bashir M, Campbell R, DeStefano L. Developing a new digital forensics curriculum. Digital Investigation 2014; 11: S76-S84. DOI: 10.1016/j.diin.2014.05.008.

[40] Lutkevich B. Computer Forensics (Cyber Forensics) (2021, May). Available at: https://searchsecurity.techtarget.com/definition/computer-forensics [accessed July 28, 2021].

[41] Williams L. What is Digital Forensics? History, Process, Types, Challenges (2021, October). Available at: https://www.guru99.com/digital-forensics.html [accessed October 23, 2021].

[42] Maheshwari A. Microsoft "Age of Intelligence" whitepaper 2019. Available at: https://news.microsoft.com/en-in/features/

microsoft-india-ai-whitepaper-age-of-intelligence/ [accessed August 31, 2021].
[43] Marr B. What are the latest trends in data science? (2021, August). Available at: https://bernardmarr.com/what-are-the-latest-trends-in-data-science/ [accessed October 31, 2021].
[44] Zatyko K. Defining Digital Forensics. Forensic Magazine 2007.
[45] Farmer D, Venema W. Forensic Discovery. 2004. Addison-Wesley. Reading, MA.
[46] Sakpal M. How to Use Citizen Data Scientists to Maximize Your D&A Strategy (2021, June). Available at: https://www.gartner.com/smarter-withgartner/how-to-use-citizen-data-scientists-to-maximize-your-da-strategy [accessed October 28, 2021].
[47] Fakiha B. Digital Forensics: Crimes and Challenges in Online Social Networks Forensics. Journal of the Arab American University 2020; 6(1): 15–35.
[48] Buhr S. An Amazon Echo may be the key to solving a murder case (2016, December). Available at: https://techcrunch.com/2016/12/27/an-amazon-echo-may-be-the-key-to-solving-a-murder-case/ [accessed October 31, 2021].
[49] Islam MR, Liu S, Wang X, Xu G. Deep learning for misinformation detection on online social networks: a survey and new perspectives. Social Network Analysis and Mining 2020; 10(1): 1–20. DOI: 10.1007/s13278-020-00696-x.
[50] Orozco AL, Huamán CQ, Álvarez DP, Villalba LJ. A machine learning forensics technique to detect post-processing in digital videos. Future Generation Computer Systems. 2020; 111: 199–212. DOI: 10.1016/j.future.2020.04.041.
[51] Smys S, Raj JS. Analysis of Deep Learning Techniques for Early Detection of Depression on Social Media Network-A Comparative Study. Journal of trends in Computer Science and Smart technology (TCSST). 2021; 3(1): 24–39. DOI: 10.36548/jtcsst.2021.1.003.

2

Mitigating and Controlling Virtual Addiction Through Web Forensics and Deep Learning

R. Danu[1], and Dr. S. Kavitha[2]

[1]Research Scholar, SRM Institute of Science and Technology, Kattankulathur, Chennai 603203.
[2]Associate Professor, SRM Institute of Science and Technology, Kattankulathur, Chennai 603203
Email: dr2402@srmist.edu.in; kavithas@srmist.edu.in

Abstract

Addiction is categorized as a brain disorder, rather than a human flaw or decision. The world is presently in the hands of digital technology, and the majority of human jobs have been delegated to machines. Virtual systems are being used to access almost every function, which leads to virtual addiction. Virtual addiction is expanding in this digital environment as smart systems mature. Deep learning has already been explored and applied in a multitude of conditions, with spectacular results. As a consequence, such a topic warrants more research in order to be beneficial for subsequent implementations. There are numerous layers between the input and the output in deep learning. A layer can include hundreds, if not thousands, of neuronal units. The layers between the input and output are the hidden layers, and the nodes are referred to as the hidden nodes. Auditing human behaviour systems on web history is necessary to mitigate virtual addiction, and a framework is provided that utilizes role-based access control (RBAC) and web forensics to track how long apps or websites are viewed regularly using a user permission agreement (UPA). The primary goal of this chapter is to deliver an overview of deep learning's use in preventing and managing virtual addiction.

2.1 Introduction

A growing percentage of children and adolescents prefer to live in a virtual world dominated by computers and televisions, ignorant of the issues in the actual world. Computers and the Internet appear to be overused at best and addictive at worst. As per a survey [1], men seem to be more susceptible to internet addiction (IA) than women. Therefore, men and women may differ in particular online consumption data, and categories and associated Internet addiction.

A highly psychoactive experience is created by combining available stimulating material, ease of access, convenience, cheap cost, visual stimulation, autonomy, and anonymity. These technologies have an impact on our daily lives and relationships. Some of these consequences are not entirely favorable and may contribute to various undesirable psychological repercussions. According to CNNIC (China Internet Network Information Centre) poll, addiction to the Internet was one of the most important mental health concerns among Chinese adolescents, with over 17% becoming addicted to the Internet in 2014 [2]. Medical practitioners and researchers use deep learning to identify hidden opportunities in data and to improve healthcare. Deep learning employs neural networks to speed up processing and give more accurate results, and may reveal previously undiscovered possibilities and trends in clinical data in healthcare. This chapter looks at how deep learning techniques may be used to overcome Internet use, which is considered a major virtual addiction.

2.2 Internet Addiction (IA) Types

Internet addiction refers to various online, computer, and mobile technology-related behaviors and impulse control issues.

2.2.1 Cyberbullying addiction

Cyberbullying addiction is one of the most personality related online addictions. It includes pornographic material, porn websites related to sex, fictional chat sessions, and pornographic video sites. Obsession with any of these services might endanger one's capacity to develop genuine sexual, romantic, or personal connections in the real world.

2.2.2 Web obligations

Web obligations are online interactive habits that may be highly harmful, including internet poker, financial services, online marketplaces, and compulsive internet ordering. These behaviors can have a detrimental impact on one's economic stability and lead to workplace issues. Overspending or financial loss can also cause stress in a person's connections. It is easy to get into online gambling sites and organizations.

2.2.3 Addiction to cyberspace relationships

Cyber or online connection junkies are preoccupied with establishing and sustaining relationships online, frequently disregarding and forgetting about real-life family and friends. Typically, online connections are developed in chat rooms or other social networking platforms.

2.2.4 Anxious searching for content

Internet users have access to a wealth of statistical information. The efficiency with which they can acquire information can turn into an obsession, with gathering and organizing data for certain people. The cognitive process can also be a sign of underlying intense disorders.

2.2.5 Gaming addiction

Either real or virtual device activities are included in the definition of internet gaming disorders. As machines grew increasingly widely available, applications like Chess, Sudoku, and Dig Dug were built into source codes.

2.2.6 Smartphone mobile app addiction

We are attracted by the videogames, programs, and digital applications that link us to our smartphones. Addiction to social networking applications, dating apps, texting, and messaging may lead to virtual, online friendships being more important than real-life connections. Social media addiction can be fueled by a Web overuse problem or behavioral addiction, often known as "nomophobia".

2.3 Human Behavior Analysis

Human behavior was once thought to be unpredictable for two reasons. Humans are pretty proud of their capacity to be unpredictable. However, as AI and deep learning evolve, as well as other technological breakthroughs, the scenario is changing.

- Humans are sophisticated creatures that do things just because they want to.

- Inability to identify and understand elements that impact human decision-making.

Machines and technology have advanced to the point that they can now anticipate human behavior and future occurrences. According to a study, more than 45% of the world's population utilizes social media (Emarsys) [3], with Facebook being the most popular platform (68% of US adults use Facebook) (Pewinternet) [3]. People are increasingly utilizing social websites for various activities other than publishing photos, such as shopping, finding the finest locations to visit or eat, sharing events, and raising awareness.

According to Figure 2.1, the degree of semantics and the length of time required in the study, the levels of human behavior analysis are divided into motion, action, activity, and behavior. Tasks such as movement detection, background extraction, and segmentation are confronted at the motion level. Human body motion is sensed and recognized at the action level to determine what a person is doing or the things with which they are engaging. As a result, human behavior analysis is carried out from this point onwards. In order to comprehend human behavior, a collection of numerous acts is categorized at the activity level. A highly semantic understanding is essential at the behavioral level.

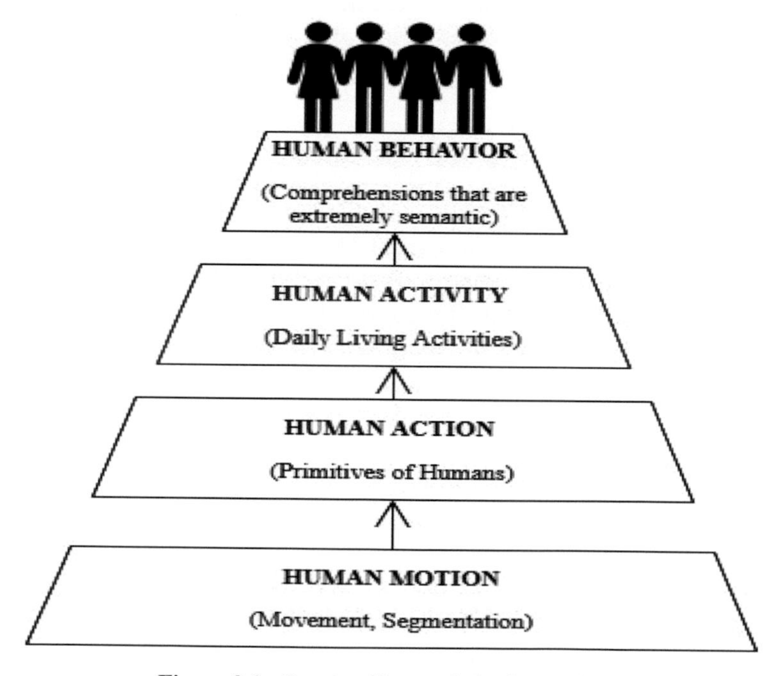

Figure 2.1 Levels of human behavior analysis.

Ways of life, personal habits, timetables, and routines may all be studied over a period of days to weeks. Abnormal behaviors and abnormalities can be noticed in this time, for example, to diagnose senile dementia early [4].

2.4 Deep Learning's Relevance to Human Behavior Prediction

Deep learning is an essential part of data science because it supports predictive analytics and allows data scientists to rapidly and cheaply gather, organize, analyze, and understand massive quantities of data. After analyzing AI and deep learning algorithms' capacity to observe and anticipate human behavior, one important element that impacts the quality of the algorithm's outputs is the number of datasets on which the deep learning models are trained. The more data there is and the higher the quality, the more accurate and trustworthy the results will be.

2.5 Forms of online mining

The forms of online mining depicted in Figure 2.2 include HTML page information extraction, keywords, standard associated metadata, and customized usage monitoring.

2.5.1 HTML page information extraction

HTML page information extraction is extracting key data from website material, metadata, or online assets.

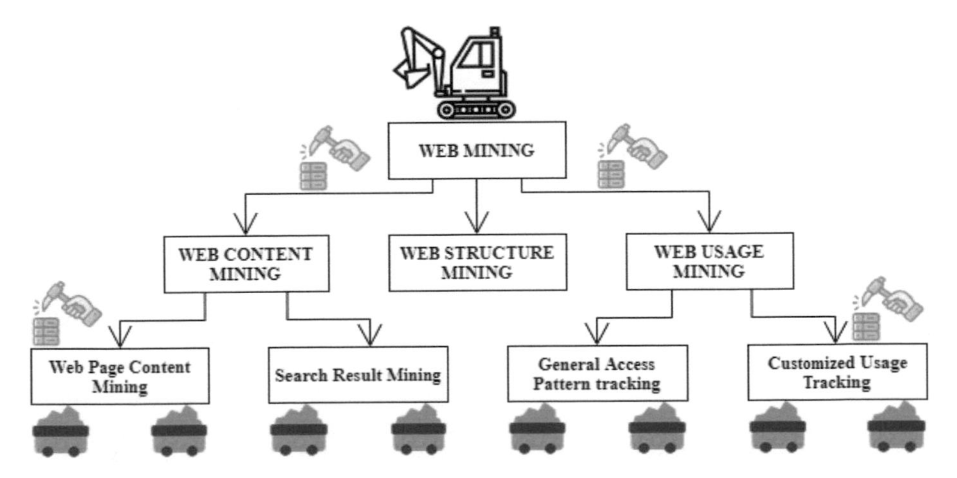

Figure 2.2 Forms of online mining [16].

2.5.2 Commonly associated metadata extraction

Commonly associated metadata extraction aims to discover the model that supports the network's link architectures. The model is based on a connection architecture, which can include or exclude connection details. This approach may classify website content and provide data such as website similarities and relationships.

2.5.3 Customized web usage monitoring

Customized web use monitoring aims to analyze the information supplied by a website surfer's activities or habits. Website information and structure mining employs actual or primary web content. In contrast, customized use tracking uses secondary collected data of the user's activities while engaging with the web. The customized usage monitoring data includes data from website server log data, browser logs, proxy server logs, login details, enrollment data, user sessions (or) payments, cookies, user queries, cache data, button presses and navigation, and any other information gleaned via interactions [7].

The various types of internet mining are outlined in Table 2.1. Information extracted from HTML sites might include content, hyperlinks, and even attributes implied or provided by the visitor. The same can be stated for commonly related metadata extraction, which may take advantage of both hyperlink and hyperlink structure information. User information is frequently used in applications such as user modeling and intelligent machines.

2.6 Web Usage Mining Process

The discovery of relevant trends and patterns is one of the primary objectives of web usage mining. Such patterns and data may frequently provide crucial information about a company's consumers or system users. Preprocessing, pattern identification and pattern analysis [9] are the three key phases in web usage mining. Preprocessing, like other data mining applications, entails data cleaning.

Technology such as web cookies and user registration have been implemented in specific applications. The generic machine learning and data mining techniques such as association rule mining, clustering and classification are frequently used in pattern identification and analysis. Web use mining apps, on the other hand, have a number of problems, one of which is that the webserver log data is anonymous, which makes it impossible to identify people and user sessions from the data.

Table 2.1 Forms of online mining [7].

	HTML page information extraction			
	Information retrieval view	**Database view**	**Commonly associated metadata extraction**	**Customized usage monitoring**
Aspects of information	✓ Structureless ✓ Generalized moderately	✓ Generalized moderately ✓ Using the Internet as a dataset	✓ The architecture of the connections	✓ Interconnectivity
The most important information	✓ Record in plaintext ✓ Record in hypermedia	✓ Record in hypermedia	✓ The architecture of the connections	✓ Server logs ✓ Browser logs
Portrayal	✓ Words and keywords ✓ Theme ✓ Collaboration	✓ OEM graph ✓ Pattern recognition	✓ Graph	✓ Relational table ✓ Graph
Procedure	✓ Natural language processing	✓ Unshared algorithms ✓ Inductive logic programming	✓ Unshared algorithms	✓ Empirical ✓ Associations Guidelines
Types of activities	✓ Classification ✓ Grouping ✓ Retrieval guidelines	✓ Identifying two or more sub-structures ✓ Determining the structure of a webpage	✓ Classification ✓ Grouping	✓ Advertising ✓ Building a model by consumers

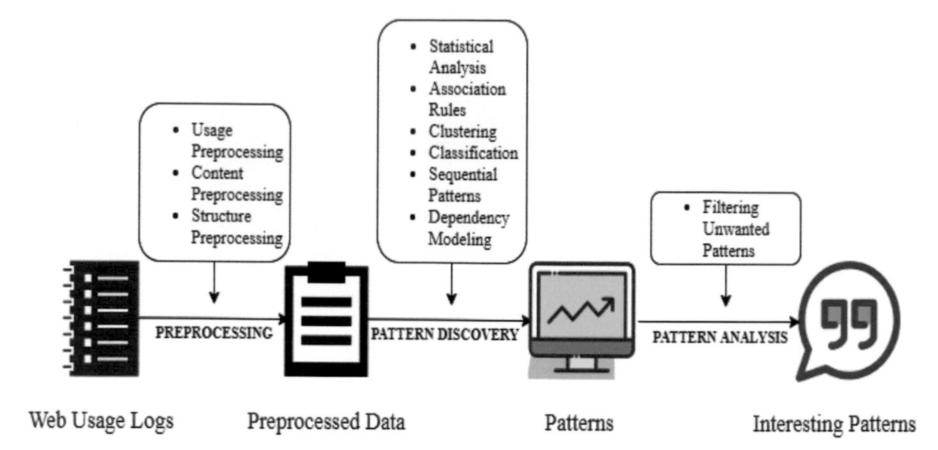

Figure 2.3 Exploration of web usage.

Table 2.2 An example of a web browser log.

#	I.P. Address	Time	URL
1		[26/May/2000:02:04:41-05012]	"GET A.html HTTP/1.0"
2		[26/May/2000:02:05:41-05012]	"GET B.html HTTP/1.0"
3		[26/May/2000:02:06:41-05012]	"GET D.html HTTP/1.0"
4		[26/May/2000:02:07:41-05012]	"GET C.html HTTP/1.0"
5		[26/May/2000:02:08:41-05012]	"GET L.html HTTP/1.0"
6		[26/May/2000:02:09:4 -05012]	"GET R.html HTTP/1.0"
7	132.466.68.8	[26/May/2000:02:10:41-05012]	"GET D.html HTTP/1.0"
8		[26/May/2000:02:11:41-05012]	"GET B.html HTTP/1.0"
9		[26/May/2000:02:12:41-05012]	"GET A.html HTTP/1.0"
10		[26/May/2000:02:13:41-05012]	"GET C.html HTTP/1.0"
11		[26/May/2000:02:14:41-05012]	"GET D.html HTTP/1.0"
12		[26/May/2000:02:15:41-05012]	"GET F.html HTTP/1.0"
13		[26/May/2000:02:16:41-05012]	"GET G.html HTTP/1.0"

Three primary tasks are doing web usage mining or web usage analysis, as indicated in Figure 2.3. Preprocessing is transforming use, content, and structural information from multiple data sources into data abstractions required for the discovery of patterns. Pattern discovery incorporates algorithms and methods from various fields, including data mining, statistics, machine learning, and pattern recognition. As shown in Figure 2.3, the pattern analysis is the final stage in the whole web usage mining process. The pattern analysis is also used to remove uninteresting rules (or) patterns from the discovered collections during the pattern discovery phase. Table 2.2 displays the results. The IP address, user id (Uid), time, URL, and agent are examples of logs from a web server. Due to the apparent growth in the popularity of web-based apps, there is growing interest in studying web usage data to understand it better and apply what we have learned to serve consumers effectively.

2.7 RNN-based Analysis of Web History Log Data

Recurrent neural networks (RNNs) are a type of neural network that may be used to represent sequence data. RNNs, derived from feedforward networks, behave in the same way as human brains. As shown in Figure 2.5, the previous step's output is supplied as input to the next step in an RNN [12]. Recurrent neural networks have a version called long short-term memory (LSTM) networks.

2.8 Feed-forward Networks Versus RNNs

The method of RNNs and the information of feed-forward neural networks channel gives them their names. As indicated in Figure 2.4, in a feed-forward neural network, data only flows in one direction from the input layer to the output layer and also passes via the hidden layers. Data travels straight through the network, never passing through the same node twice. The feed-forward neural networks have no recollection of the information. They receive poor predictors of what will happen next. Because a feed-forward network only examines the current input, it is called a "feed-forward" network.

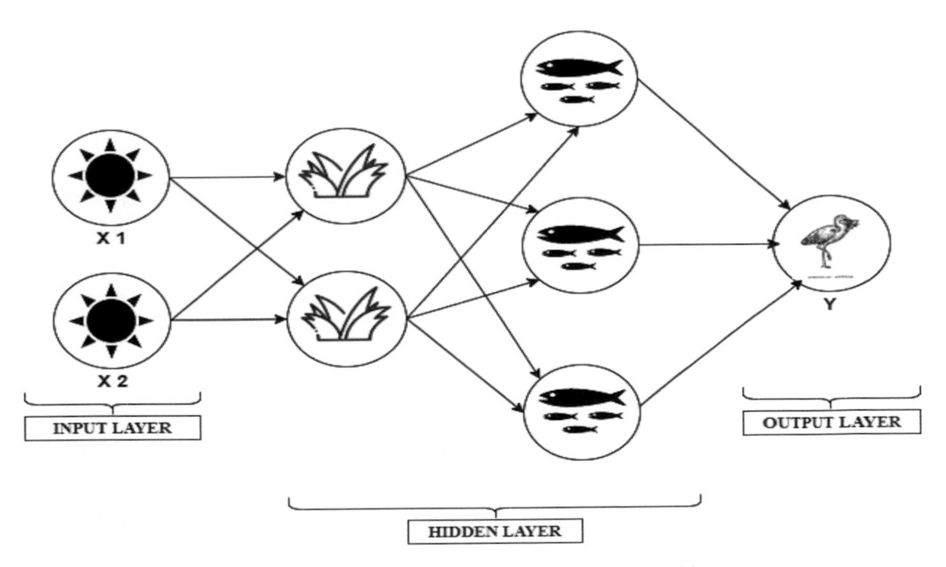

Figure 2.4 Feed-forward neural network basic architecture.

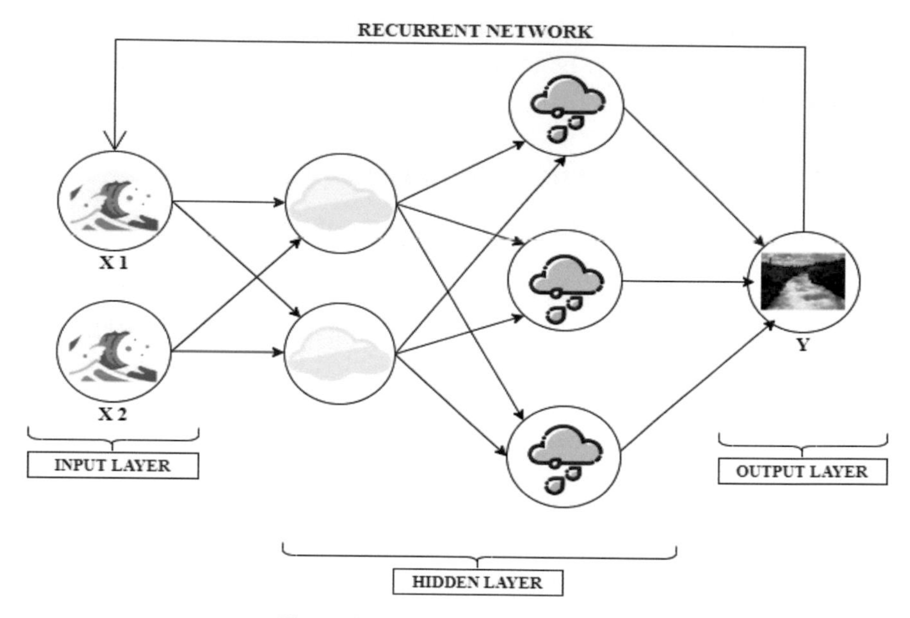

Figure 2.5 RNN basic architecture.

The information is an RNN cycle via a loop. When it makes a judgment, it considers the current input and what it has learnt from the prior inputs. Short-term memory is also present in a typical RNN. They also have long-term memory when used in conjunction with an LSTM. As shown in Figure 2.5, the present and recent past are the two inputs of the RNN. This is significant because the data sequence provides critical information about what will happen next, so an RNN can perform tasks that other algorithms cannot. Like all other deep learning algorithms, the feed-forward neural network gives a weight matrix to its inputs before producing output. The RNNs map the one to many, many to many (translation), and also many to one, whereas feed-forward neural networks map the one input to the one output (classifying a voice) [13].

2.9 RNN Relying on LSTM

The LSTM algorithm extracts context-dependent information from sequential data, such as information about text inputs (or) video data. It is a significant upgrade from standard RNNs. It can learn long term dependencies efficiently, which is not feasible with standard RNNs. Because people also apply contextual knowledge to anticipate future data, LSTMs become substantially more mortal. LSTMs can retain prior knowledge over a lengthy period, which makes them helpful in various applications. LSTMs are designed to work

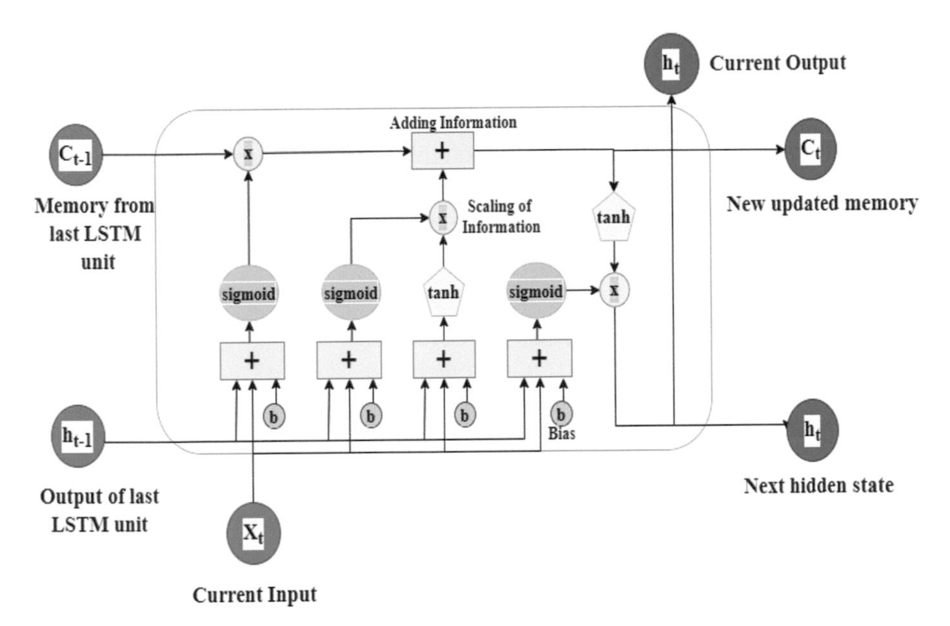

Figure 2.6 LSTM in-depth architecture [13].

with long sequences of data with up to 400 time steps [11]. The advantage of LSTM for sequence classification is that it can also learn directly from the raw time-series data, eliminating the need for domain knowledge and manually building input characteristics. The model should be able to learn an internal representation of the time series data and perform similarly to models trained on a dataset version with artificial features.

A standard LSTM unit is shown in Figure 2.6, composed of a cell-like input, output and forget gates. The values in the cell are remembered over arbitrary time intervals, and three gates regulate the information flow into and out of the cell. Both cell state and hidden state are transmitted to the subsequent cell. The main chain of data flow is the cell state which permits the data to pass forward practically unaltered. There may, however, be some linear transformations. Sigmoid gates allow data to be added to or deleted from the cell state. A gate is similar to a layer or a set of matrix operations with weights. Because it utilizes gates to regulate the memorization process, LSTMs are meant to avoid the long term reliance problem. First the information stage needs to be identified in building an LSTM network that is not needed and will be left out of the cell.

The sigmoid function takes the last LSTM unit (h_{t1}) output at the time (t_1) and the current input (X_t) at the time (t), determining the recognizing process and excluding data. The sigmoid function decides whether parts of the output (old) should be removed. The forget gate (f_t) is a gate in which it is a vector with ranging values from 0 to 1 [13] that corresponds to each number in the cell state (C_t). The layer of sigmoid decides whether new information should be updated (or) ignored, and the tanh function gives the weight to the values which passed by deciding their importance level (−1 to 1). The two values are multiplied and updated in the new cell state. Then this new memory is added to old memory (C_{t-1}), resulting in C_t.

2.10 Various Categories of Forensics

After an issue has emerged, the forensics method starts. Figure 2.7 involves a set of established procedures to ascertain the evidence source.

2.10.1 Digital forensics

Digital forensics is the technology process towards the recognition, collection, inspection, and assessment of information while retaining the data's quality and consistent traceability.

2.10.2 Forensics over networking

The evidence from the network is identified and analyzed by network forensics. It obtains information about the access points that have been used to extract content.

2.10.3 Web forensics

Evidence is found in the subscriber's cache, short system logs, database, conversations, session records, and caches when using website analytics.

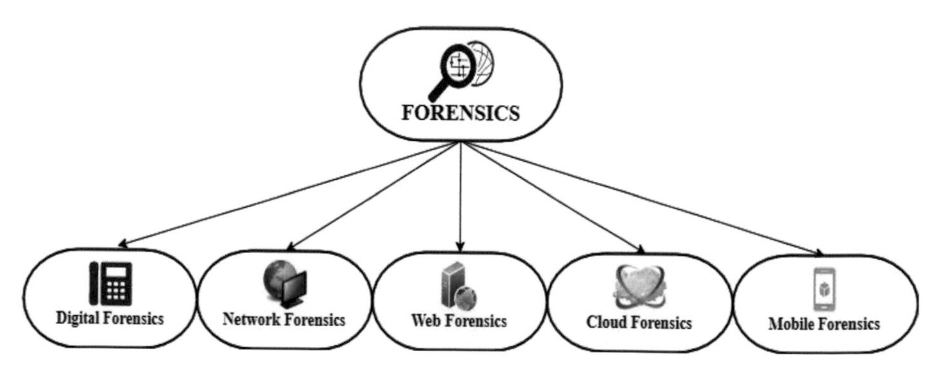

Figure 2.7 Categories of forensics.

2.10.4 Cloud forensics

Cloud forensics is a type of internet forensics that involves computer forensics cloud-based services. Some examples include log files, user activity, secure authentication logs, DBMS logs, and other information resources.

2.10.5 Mobile forensics

Cellphone forensics is a computer forensics branch that also identifies the evidence from portable devices. Conversation records, text messages, and the device's storage are all used to compile the proof.

2.10.6 Web browser forensics

Since many criminal and civil cases may be founded on evidence obtained from user online activity, web browser forensics is a key element of computer forensics. Web browser forensics is most commonly used to check for previously seen material by monitoring things like search history and regular internet usage on some kind of personal computer or mobile phone. This also involves examining log files from a server to gain correct information about a specific machine and tracking activity on a website. Viewing history, favorites, downloads, cookies, and the cache is where an inspector might look for evidence in a browser [15].

2.11 Web Browser Artifacts

Each browser saves its files in a different location and gives them various names, but they all save the same sort of data (artifacts) most of the time. Let us look at the most frequent artifacts that browsers save.

2.11.1 Navigation history

The user's navigation history is stored in the navigation history field. It can be used to see if the user visits certain websites repeatedly.

2.11.2 Autocomplete data

The data that the browser suggests depending on what you search for the most, is known as autocomplete data. It may be utilized in conjunction with the navigation history to acquire additional perspective.

2.11.3 Cache

The browser produces various cache data (images, javascript files, and so on) when visiting websites to speed up the loading time. During a forensic investigation, these cache files might be a valuable source of data.

2.11.4 Favicons

Favicons are tiny icons that may be found in tabs, URLs, and bookmarks. They can be used to learn more about the website or locations the user has visited.

2.11.5 Browser session storage

A typical method for preserving data in a browser is session storage, allowing programmers to save and retrieve multiple variables. It only keeps data for one session at a time, and the data is erased when the browser window is closed.

2.11.6 Form data

Anything entered into forms is frequently saved by the browser, allowing the browser to recommend previously submitted data when the user fills out a form.

2.12 Analysis of Website Usage History

Website usage history analysis aims to extract useful information or knowledge from log files depending on the key sets of variables used in the extraction and processing. Several websites apply website usage history studies to identify consumer data and performance in terms of the website's strengths and shortcomings. The results of weblog mining can be used for various purposes, including personalizing web content delivery and improving user mobility through prefetching and buffering, which attracts more customers [6]. Human behavior can also be analyzed, and we can determine the level of virtual addiction of a particular human. According to Figure 2.8, the following tasks make up the breakdown of weblog mining [7]:

- The initial step is resource discovery, which entails retrieving the desired online content.

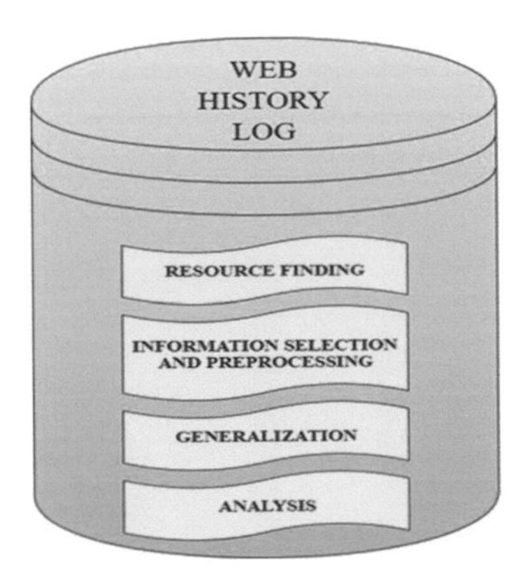

Figure 2.8 Analysis of web history log.

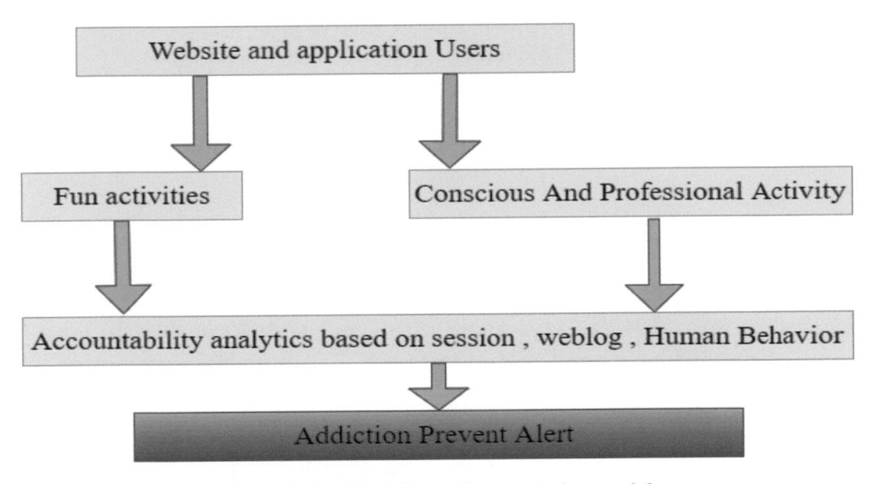

Figure 2.9 Workflow of an analytics model.

- The following task is information selection and preparation, which is the act of automatically picking and preparing specific data from a web source.
- General patterns are discovered automatically at individual websites and across various websites via generalization.
- Finally, the validity and interpretation of the mined pattern are examined.

Figure 2.9 depict the workflow of an analytics model. The graphical analysis of web usage is shown in Figure 2.10. The x-axis represents the URL, and the y-axis represents the number of times the user visits a particular URL. Based

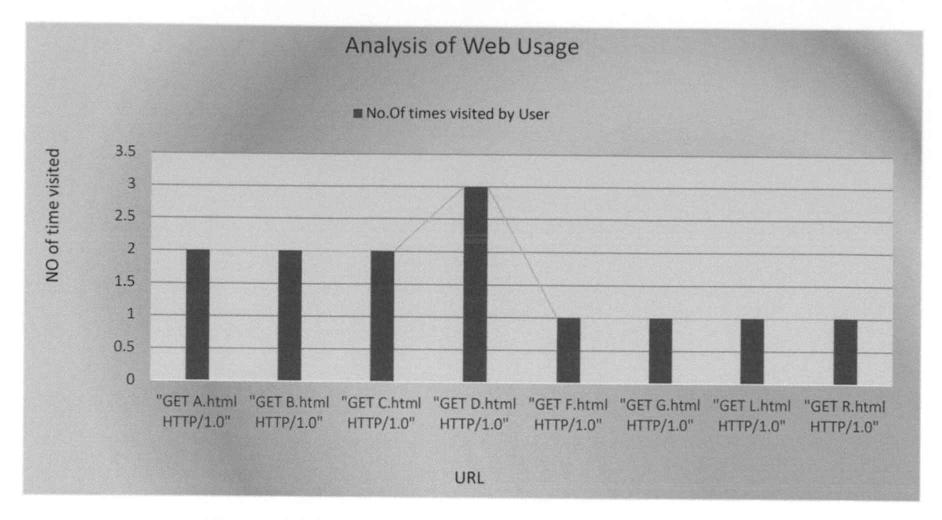

Figure 2.10 Graphical analysis of a web browser log.

on the analysis below, we learned that the URL "GET D.html HTTP/1.0" is frequently visited by the user via the I.P. address "132.466.68.8."

2.13 Conclusion

This chapter aims to give an updated overview of the constantly expanding field of customized web usage monitoring and web forensics. In addition, this chapter clarifies several misunderstandings about the concept of "web mining." Moreover, three web mining categories have been defined, and their relationships have been analyzed. Exploration of online usage data and analysis of user activities and preferences are significant parts of this chapter; deep learning technologies are used to improve human behavior detection abilities. The primary goal of using deep learning to analyze web usage data is to understand web usage better and implement it to avoid virtual addiction. The LSTM neural network is one of the most advanced designs for dealing with sequential data or time series problems. LSTM also learns the long-term connections between the successive datasets and demonstrates consistent effectiveness in virtual addiction diagnosis.

2.14 Acknowledgement

I would like to convey my gratitude to my friends for the stimulating talks. I thank Dr. S. Kavitha, Associate Professor, Department of Computing Technology, SRM Institute of Science and Technology, for assisting me in finishing the chapter. Last, but not least, I want to express my gratitude to my institute, SRM Institute of Science and Technology, for providing me with this beautiful chance to work in such a wonderful atmosphere.

References

[1] "Do men become addicted to internet gaming and women to social media? A meta-analysis examining gender-related differences in specific internet addiction" [W. Su et al. Computers in Human Behavior 113 (2020)].
[2] "Parenting styles and internet addiction in Chinese adolescents: Conscientiousness as a mediator and teacher support as a moderator" [R.-p. Zhang, et al. Computers in Human Behavior 101 (2019) 144–150].
[3] https://www.sganalytics.com/blog/ai-and-deep-learning-to-predict-humanbehaviours-can-the-unpredictable-be-predicted.
[4] https://www.researchgate.net/publication/259842872_Vision_based_Recognition_of_Human_Behaviour_for_Intelligent_Environments Alexandros_Andre_Chaaraoui.
[5] "Human Behavior Recognition Using Deep Learning" [Jia Lu, Wei Qi Yan and Minh Nguyen. 978-1-5386-9294-3/18/$31.00 ©2018 IEEE]
[6] "Extracting Knowledge from Web Server Logs Using Web Usage Mining" [Mirghani. A. Eltahir, Anour F.A. Dafa-Alla, 2013 International Conference on Computing, Electrical and Electronic Engineering (icceee), 978-1-4673-6232-0/13/$31.00 ©2013 IEEE].
[7] Web Mining Research: A Survey [Raymond Kosala, Hendrik Blockeel, july 2000 vol2, issue 1, SIGKDD].
[8] https://citeseerx.ist.psu.edu/viewdoc/download?doi=10.1.1.83.4596&rep=rep1&type=pdf.
[9] [Srivastava, J., Cooley, R., Deshpande, M., & Tan, P. N. (2000).] Web usage mining: Discovery and applications of Web usage patterns from Web data. ACM SIGKDD Explorations, 1(2), 12–23.
[10] https://builtin.com/data-science/recurrent-neural-networks-and-lstm"A Guide to RNN: Understanding Recurrent Neural Networks and LSTM Networks".
[11] https://machinelearningmastery.com/how-to-develop-rnn-models-for-human-activity-recognition-time-series-classification/.
[12] "Web Page Categorization using RNN Based on URL" [Pratiksha Vaishnav, and Ankit Kalariya International Journal of Advanced Research in Science, Communication and Technology (IJARSCT) Volume 6, Issue 1, June 2021].

[13] "Application of Long Short-Term Memory (LSTM) Neural Network for Flood Forecasting" [Xuan-Hien Le, Hung Viet Ho, Giha Lee, and Sungho Jung] Water 2019, 11, 1387; doi:10.3390/w11071387.

[14] "Data Collection Techniques for Forensic Investigation in Cloud"DOI: 10.5772/intechopen.82013 https://www.intechopen.com/chapters/64377.

[15] "Web Browser Forensics for Detecting User Activities" Mayur Rajendra Jadhav, Dr. Bandu Baburao Meshram, International Research Journal of Engineering and Technology (IRJET), Volume: 05, Issue: 07, July 2018.

[16] "Web mining and Web usage mining techniques", Nasrin Jokar, Ali Reza Honarvar, Shima Aghamirzadeh, Khadijeh Esfandiari, Bulletin de la Société des Sciences de Liège, Vol. 85, 2016, p. 321–328.

[17] Sabapathi, V., & Selvin Paul Peter, J. (2021). Classification of Addiction Behavior based on Regular and Rare Model. Journal of Scientific and Industrial Research (JSIR), 80(7), 593–599.

[18] Sabapathi, V., & Vijayakumar, K. P. (2020). A Study of Addiction Behavior for Smart Psychological Health Care System. Role of Edge Analytics in Sustainable Smart City Development: Challenges and Solutions, 257–272.

[19] "A novel text mining approach for scholar information extraction from web content in Chinese" Xia Xie, Yu Fu, Hai Jin, Yaliang Zhao, Wenzhi Cao, Elsevier, Future Generation Computer Systems 111 (2020) 859–872.

[20] "Assisted pattern mining for discovering interactive behaviors on the web", A. Apaolaza and M. Vigo, International Journal of Human-Computer Studies 130 (2019) 196–208.

[21] "Mining meaningful and rare roles from web application usage patterns", Nurit Gal-Oz, Yaron Gonen, Ehud Gudes, computers & security 8 2 (2019) 296–313.

[22] "Development of a web-platform for mining applications", C. Newman et al. International Journal of Mining Science and Technology 28 (2018) 95–99,1569-190X 2017 Elsevier B.V.

[23] "A new web-based solution for modelling data mining processes" V. Medvedev et al. Simulation Modelling Practice and Theory 76 (2017) 34–46.

[24] "Do men become addicted to internet gaming and women to social media? A meta-analysis examining gender-related differences in specific internet addiction", W. Su et al, Computers in Human Behavior 113 (2020) 106480.

[25] "Internet addiction at workplace and it implication for workers life style: Exploration from Southern India", A. Shrivastava et al, Asian Journal of Psychiatry 32 (2018) 151–155.

[26] "Internet use behaviors, internet addiction and psychological distress among medical college students: A multi Centre study from South India" N. Anand et al, Asian Journal of Psychiatry 37 (2018) 71–77.

3

Automatic Identification of Cyber Predators Using Text Analytics and Machine Learning

N. Kavitha[1], K. Ruba Soundar[2], S. Shanmuga Priya[3], and T. SathisKumar[4]

[1]Indra Ganesan College of Engineering, Trichy, Tamil Nadu
[2]Mepco Schlenk Engineering College, Sivakasi, Tamil Nadu
[3]New Horizon College of Engineering, Bengaluru, Karnataka
[4]Saranathan College of Engineering, Trichy, Tamil Nadu
Email: nkavithamail@gmail.com; rubasoundar@gmail.com; priya.soundararajan@gmail.com; sathistrichy22@gmail.com

Abstract

Supplying safe surroundings for women and children in cyberspace, i.e., online communal groups, is considered an essential aspect of communal welfare. Due to the frequency of online exchanges, getting rid of juvenile exploitation in cyberspace has become necessary. A productive and extendable data mining technique is essential to solve this type of crime. The issue can be portrayed as textual pre-processing in data mining and a twofold tier architecture will assign the work based on work criteria measures. That measure is represented by the term scholarship. Exchange logs are instructive digital vestiges available on social media platforms (SMPs) that can be used to discover and identify prejudicial gesture against the cybercrimes targeting children. They serve as an essential tool to protect children from being abused by cyber predators. The proposed method employs automatic manual analytics implemented by machine learning (ML) to enable spontaneous identification of damaging text in conversation logs. Machine learning has already been successfully applied in the area of manual analysis to discover cyber sexual predaceous conversations. Notwithstanding, no literature has shown how ML can contribute to a digital forensic examination. So, the benefit of our paper is that it works out the tasks in a digital forensic examination that can be planned so as to obtain useable ML results when delving into online predators.

3.1 Introduction

Online cyber predators are an increasing risk to women and children who increasingly using communal networking and text services to communicate with unknown persons. Madigan et al., 2018 found that one out of 10 teenagers received unwanted online solicitations. The use of a child's social media can be monitored by their parents, which is not necessarily a feasible task due to the availability of a variety of communal network services and the number of ways of using them. Monitoring can be automated by installing several software tools to observe children's online behaviour. A few examples are [2, 3, 4] FlexiSPY, 2019, Easemon Inc., 2019 and Cocospy, 2019. However, these monitoring tools examine the recorded data and give an "alert" when an unwanted word is texted, which becomes a static way of analysing the child's behaviour. These tools do not use natural language processing, which may help understand the semantics of the conversation, and increases the parents' burden to identify the cyber predator manually. So applying natural language processing in any automated method could be of tremendous benefit.

The definition of a predator in terms of psychological aspect has two significant components [11]: *age inequality* and *inappropriate relationship*. The former depends on the law enforced in various countries with the psychosomatic level of the child compared to that of an adult (expected to be a predator). The latter involves the attitude of an adult to make a relationship with the child that usually involves hidden and open erotic comments. Olson, Daggs, Ellevold, & Rogers, 2007 presented a fundamental psychosomatic theory to identify the predators in online communication. They first tried to gain access to the victim, lured and groomed to make the victim accept erotic advances, and finally maintained the offensive relationship.

Online platforms act as the sources for the predators to identify victims. After that, they are distinguished by the surveillance of the predator's attempt to desensitize the child to the inappropriate relationship. Finally, the third step involves explicit erotic exploitation of the child. At this point, there is a need for a reliable online predator identification (OPI) system to enforce the law against the predator from approaching the victim.

3.1.1 OPI problem definition

Text analytic algorithms are proficient in recognizing the predators to enforce the law and have been used for automated online predator identification over

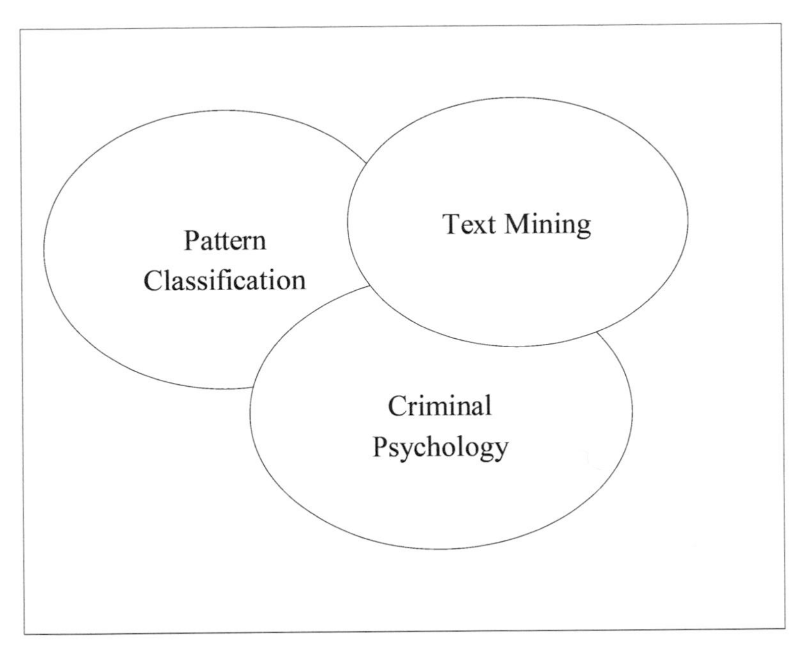

Figure 3.1 Relationship between OPI, text mining, pattern classification, and criminal psychological.

the past few decades. These algorithms are also capable of handling large volumes of chat logs. The two significant issues to be addressed by these text mining algorithms are:

1. Identify predators
2. Traverse and analyse predator illegal activity.

To identify predators, the chats can be pre-processed, and their patterns are classified to find out the psychology of the criminals (Figure 3.1). To recognize predator activity, graph mining techniques can be used.

The literature [28, 29] provides a detailed survey of the combination of text mining and pattern classification algorithms used for predator identification. The texts in the chats are analysed, and patterns are classified on the basis of all OPI, as shown in Figure 3.1. Figure 3.2 shows an illustrated version of the various techniques used in OPI and their associations.

This chapter presents a machine learning approach to traversing the communications of the cyber predator with a child through artificial neural networks (ANNs), where the approximation of activation functions occur theoretically with very few assumptions. Even though there are several practical issues in using an ANN to differentiate predatory and non-predatory conversations, it is used as the first level classifier. The following issues have

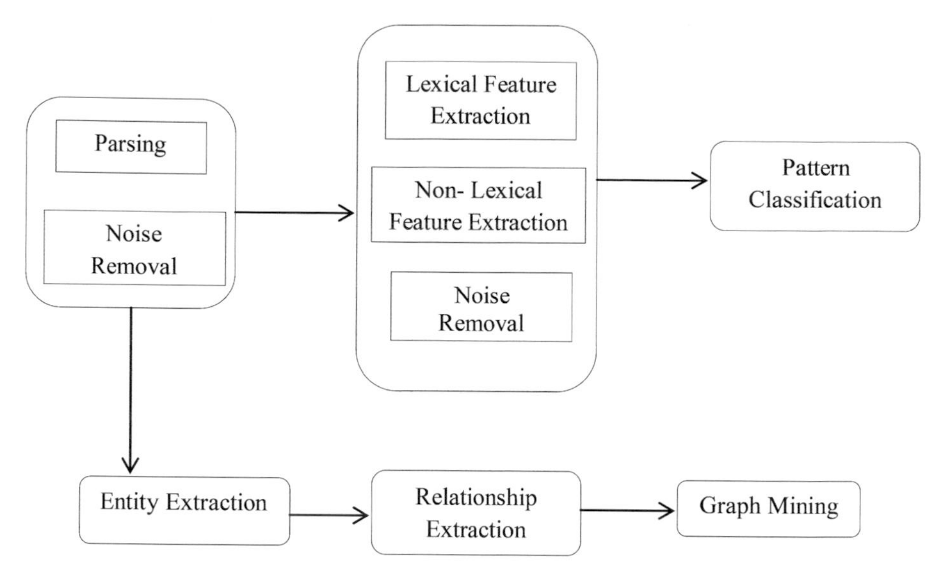

Figure 3.2 Data mining techniques used in OPI and their relationships.

to be addressed by the ANN. The problem is to significantly identify the meaning of the discussion by investigating the statements individually, which will usually generate reduced results because sentences in conversation have to be clearly interpreted within the whole context of the discussion. The relationship in prolonged discussion requires that the user input to the analysis should be an increased block of delivered sentences, which have to be analysed as a whole. One more problem to be solved is the increased proportion of the conversation, capturing thousands of texts.

The solution for the increased proportion of the conversation is provided by separating the problem into two phases. The first phase organizes the meaning of the separable texts, and it is pipelined to the second phase for classifying the whole discussion. A summary of the separable texts is passed from the first phase to the second phase to support the efficient classification of the whole discussion. A framework of five texts is used in the first phase which includes the local meaning of the individual messages. Then that information is passed to the second phase, which in turn checks the entire conversation for any texts uttered by the potential attacker.

3.2 Literature Survey

Almost all existing approaches employ deep learning techniques to detect predatory text. The existing research started with the one-to-one transcriptions created by a group of volunteers of the PJ Foundation 2019 [5] who pretended to be victims in chat rooms in order to trap predators. Their research helped to reveal around 630 cyber predators, and the conversations with predators have been recorded and made available as transcriptions to the research community as a database. Several studies explored the database [6] for its linguistic properties. Inches and Crestani, 2012b [8] and Inches and Crestani, 2012c [9] described the detail in the International Competition for Sexual Predator Identification PAN 2012 workshop, catalysing awareness of the problem where the PAN 2012 transcription was generated using the PJ dataset, Omegle chats and IRC logs modified as a person-to-person stimulating chat

In the literature, it is found that authors have used various machine learning techniques. Support vector machines [7, 11, 12, 13, 14, 15, 16, and 17] have produced a reasonable result for identifying predators. Similarly, decision trees [10, 14, 15, 16, 17, 18, 19, 20, and 22], k-nearest neighbour, logistic regression, maximum entropy, and multilayer perceptron (MLP) neural networks have produced reasonable results to classify the predators from normal persons. A rule-based heuristic has been presented [19, 20] that works better than the supervised learning approach. The methods discussed in the literature need a detailed description of the features representing the chats. The above methods have followed the vocabulary structures of single and dual entities in speech usually implemented by attackers.

The collection of uttered texts by the predator is the source of forming the vocabularies that are anticipated to indicate the chat intention. A few samples for vocabulary structures comprise the count of unaware verbs (e.g. kiss, suck) and restructured verbs (e.g. impart, rehearsal) [19, 20]. Other authors [21] have used philological investigation and word count (PIWC) structures that link words with intellectual and sensitive states. A few approaches use chatty/communicative structures that depict the prototype properties of the entire conversation, such as the total number of members involved in the chat, the duration of the chat [22], the number of the first initiations by the predator, and the questions asked by the predator [23].

Many authors have applied MLP neural networks to identify predators [14, 15, and 17] since they do not require an explicit set of structures. In turn, they have used a representation of a set of words that consolidates a chat as the total number of occurrences of every word in the language without considering their sequence order.

3.2.1 Cyber predator intent classification

The aim of the conversation of a predator with a child through online social media is classified into five groups by O'Connell [24]: forming a friendship, extending friendship to a relationship, threat assessment, uniqueness, and finally erotic behaviour. The classification is reduced to three groups by Olson [25], including mentoring, segregation, and tactic. Researchers [20] have presented a classification of cyber predator tool named ChatCoder2 that practices a heuristic approach on rules to categorise the conversation based on their grouping. The words are positioned in any of the below-mentioned categories. We have also followed the same methodology to generate a labelled dataset for training.

Acquiring individual facts through conversation (I): This category may involve queries about individual information like age, gender, place of residence or work, friends, and interests. This is the key for the cyber predator to initiate a trusting association with the child.

Mentoring (II): This category uses erotic terminology, irrespective of the conversation. Cyber predators are usually experienced in pacifying the child in erotic conversations.

Attitude (III): This category explains how the cyber predator will collect information to call the victim for a get-together or make them maintain the secrecy of the relationship from others.

Non-predatory (IV): This category includes negative messages that do not fall under any of the above categories.

3.3 System Architecture

The proposed approach is illustrated in Figure 3.3, divided into two phases. The initial phase is "chat category", which categorises every chat from the possible predator with its intent. The next phase is "chat classification" to assess the complete chat content from the potential attacker to determine if the expected attacker is a predator or not. It takes the labels from the previous phase and does not analyse the entire chat sequence, reducing processing time.

3.3.1 Chat category

During the chat category phase, every text delivered from the expected attacker in a conversation is categorized by the intention of the conversation.

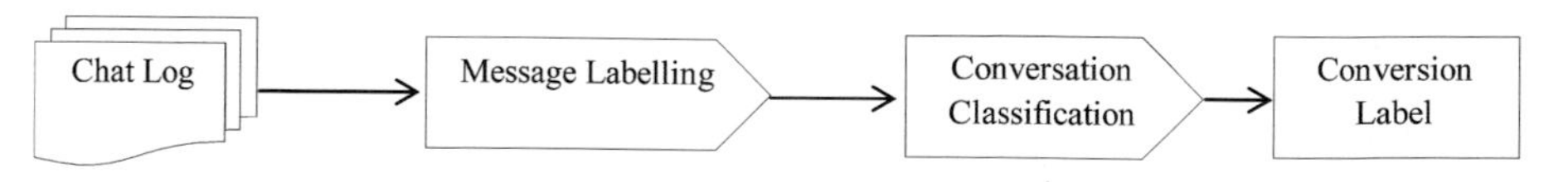

Figure 3.3 Overall structure.

The mapping pattern of the chat and the categories are trained in this stage. To precisely understand the meaning of the chat, there is a need to analyse the chat's words and the preceding chats of the conversation. Since we are processing natural language, all the chats must be converted into a vector form suitable for training. In this stage, one approach is to represent the meaning of the message and another approach is to correlate the impact of the context of the message.

The proposed approach employs Universal Sentence Encoding using TensorFlow for accomplishing the correlation between every word using the long short-term memory (LSTM) models. A message vector is generated through the chat encodings through the Universal Sentence Encoder (Cer et al., 2018a). The methods (Mikolov et al., 2013) can capture the meaning of the individual words but miss the sense in which the information is conveyed since it does not consider the sequence of the words in the text.

After encoding, the message's meaning is decided not only by the current words but also by the situation that leads it. Even though the context has similar statements, they might vary in their meanings. For example, the text "Call me" could be a chat between any intimate associates. In another case, it could be one that an attacker may use to call him or her. This is a tedious situation to find the exact meaning or the person's intention with that text alone. When classifying a message, the proposed approach considers the previous four texts in the window, leading the text into consideration, i.e. a window of five messages.

The flow of the proposed training model for the chat category is pictorially depicted in Figure 3.4. The network input layer takes input as a window

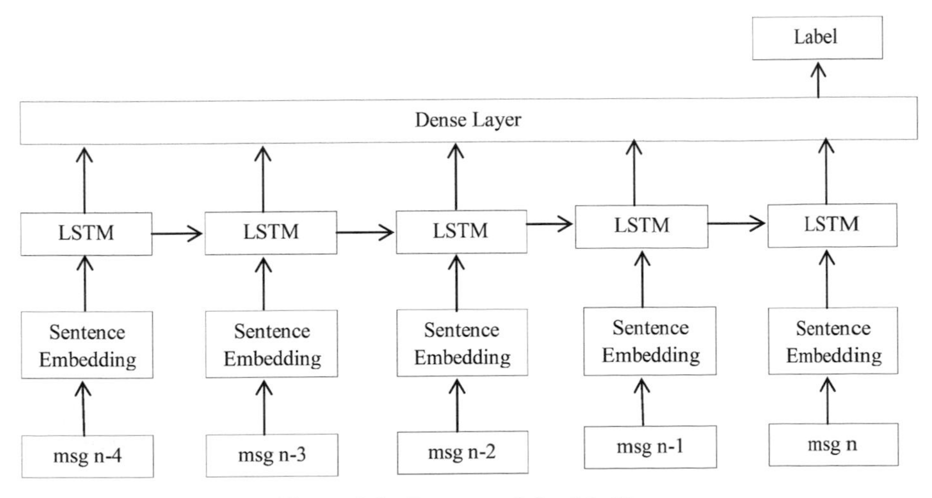

Figure 3.4 Structure of chat labelling.

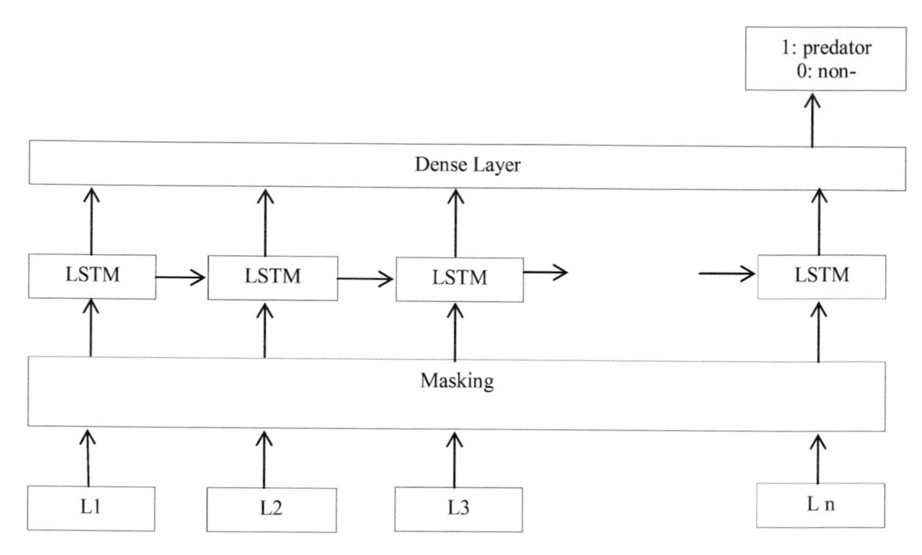

Figure 3.5 Structure of chat conversation.

of five messages, along with the message being classified at the end, for generating a 512 vector input. They form the input of the consequent LSTM followed by the dense layer. The vector window of size five at the LSTM and dense layers can conclude the native context in the proposed approach.

3.3.2 Chat classification

The first phase produces labels for the chat categories, which produces a sequence of labelled/categorized statements. The labels of the statements are used as input for the chat classification stage. The flow diagram of the second phase, chat classification, is shown in Figure 3.5. The order of the chat labels is appended to make a uniform length for the input vector in the overall chat logs. However, the appended labels are not considered as the labels for the chats. There is a mask layer in our network to ignore the padded labels. Also, the proposed approach follows training and classification of the conversation through LSTM and a dense layer.

3.4 Experiments

The proposed method presents two phases of consequences. The first phase evaluates the chat labelling stage through its precision and recall of the process. The final phase evaluates through precision and recall of both the labelling and classification stages for a set of conversations.

Table 3.1 Dataset used to evaluate chat category (phase 1).

Dataset used	Number
No. of chats considered	120
No. of total discussions	5008
No. of discussions in category I	3120
No. of discussions in category II	626
No. of discussions in category III	626
No. of discussions in category IV	626

3.4.1 Dataset

For the first set of results to evaluate the message labelling stage, the proposed method uses chatlog data from both ChatCoder2 [20] and PAN2012 [8, 9]. ChatCoder2 provides chats from the Perverted-Justice (PJ) website [5]. The ChatCoder2 dataset has all predacious conversations, while the PAN2012 dataset combines predacious and non-predacious conversations. ChatCoder2 is an investigative tool that automatically labels every message with its intent classification ('I', 'II', 'III', and IV'). The ChatCoder2 is automatically used to classify the messages in every conversation automatically. Table 3.1 describes the set of messages used to evaluate the chat labelling stage. We have taken around 5000 messages for training and testing from both databases.

The goal of the proposed method is to use the equal numeric of messages in all categories of predatory intent (II, III, and IV), i.e. around 600 in every type of message from the ChatCoder2 dataset. For balance, the same number of messages was selected with non-predacious intent (I) from the ChatCoder2 dataset. The number was selected to maintain the balance in each intent, and it is also the significant count of messages that we are left with. This number is also the minimum count of messages in each class in the ChatCoder2 dataset. 2504 (626 × 4) messages were selected from the ChatCoder2 dataset. The proposed method is considered to be a biased classification since all true negative sentences would be taken from true positive conversations from the ChatCoder2 dataset. For this reason, we selected another set of messages of around 2000 non-predatory messages from the other dataset as well.

For evaluation, the PAN2012 dataset is used for training and testing. The conversations considered should have at least 150 messages. The chat labelling produces an order of labelled which is pipelined as input for chat classification. Table 3.2 describes the features of the dataset to evaluate our classification.

Table 3.2 Dataset used to evaluate chat classification (phase 2).

Total dataset	Number
No. of discussions	475
No. of predatory discussions	130
No. of non-predatory discussions	350
No. of chats considered for evaluation	78120

Table 3.3 Performance of chat labelling (phase 1).

	Label	I	II	III	IV
Training	**Precision**	0.931	0.764	0.735	0.685
	Recall	0.912	0.782	0.799	0.658
Testing	**Precision**	0.916	0.778	0.734	0.681
	Recall	0.926	0.781	0.791	0.648

Table 3.4 Number of PAN 2012 messages under each category.

Label	I	II	III	IV
Total chats	59142	7060	6262	5666

3.4.2 Results of phase 1: Chat labelling

The first phase, chat labelling, segregates every text and provides a label to identify whether it is an erotic chat. The performance metrics for each chat label is shown in Table 3.3. The model has been built on an Intel Xeon CPU with a frequency of 2.3 GHz and a Tesla K80 within Google Colab. The neural network used 15 epochs to train the model with 80-20 training and testing datasets. The time to train the model was about a minute, and half a minute is required for sentence embedding.

3.4.3 Results of phase 2: Chat classification

In phase 2: chat classification, the evaluation is based on phase 1: chat labelling. From the training model, the numerical value for each category message in the PAN 212 dataset is shown in Table 3.4. It shows the dissemination of various chat labels in both positive (predatory) and negative messages (non-predatory).

The upper limit for the training sequence for the chat classification is fixed as 200, and any input less than that is appended with wild card characters. The neural network runs 15 epochs with a batch size of 32. The proposed approach follows an 80-20 rule for the training and test model, which includes a count of 384 and 96 messages from the datasets, respectively.

Table 3.5 Performance of categorization conversation.

	Actual	
Predicted	**Predatory**	**Non-predatory**
Predatory	29	3
Non-predatory	2	62

The performance metrics of the proposed model are shown in Table 3.5. The model has produced excellent precision; the F1 score and F0.5 score are 0.9. The performance of the proposed model is compared with the existing approaches [8, 9, and 20] by using the PAN2012 and ChatCoder2 datasets. The performance metrics recall, F1 score and F0.5 score of 16 other competitors are compared with the proposed model. Our method has an excellent recall value and F1 score, which helps identify the predator at the earliest stage. Also, the F0.5 score is less when compared to other methods, but the small number of false alarms is admissible instead of missing any predators.

3.5 Conclusions

The proposed method presents a way to detect predatory conversations to classify individual messages followed by entire conversations. The proposed approach first segregates the individual messages, labels them, and then classifies the entire text for detecting predatory conversations. The RNNs use the messages in the ChatCoder2 database for training and existing prelabelled chats. The proposed approach uses a window of five messages for labelling, including the previous four messages, to classify a chat. The results show that the proposed method performs far better than previous methods for recall value, which indicates even a fraction of predators that might go undetected.

References

[1] Madigan, S., Villani, V., Azzopardi, C., Laut, D., Smith, T., Temple, J. R., Browne, D., and Dimitropoulos, G. (2018). The prevalence of unwanted online sexual exposure and solicitation among youth: A meta-analysis. Journal of Adolescent Health, 63(2):133–141.

[2] FlexiSPY. (2019). FlexiSPY. https://www.flexispy.com.

[3] Easemon Inc. (2019). iKeyMonitor. https://ikeymonitor.com.

[4] Cocospy. (2019). Cocospy. https://www.cocospy.com.

[5] Perverted Justice Foundation. (2019). Perverted Justice. www.perverted-justice.com. Accessed: 2019-11-08.

[6] Black, P. J., Wollis, M. A., Woodworth, M., and Hancock, J. T. (2015). A linguistic analysis of grooming strategies of online child sex offenders: Implications for our understanding of predatory sexual behavior in an increasingly computer-mediated world. Child abuse & neglect, 44.
[7] Chiu, M. M., Seigfried-Spellar, K. C., and Ringenberg, T. R. (2018). Exploring detection of contact vs. fantasy online sexual offenders in chats with minors: Statistical discourse analysis of self-disclosure and emotion words. Child abuse & neglect, 81.
[8] Inches, G. and Crestani, F. (2012b). Overview of the international sexual predator identification competition at pan-2012. In CLEF (Online Working Notes/Labs/Workshop).
[9] Inches, G. and Crestani, F. (2012c). Overview of the international sexual predator identification competition at pan-2012. In CLEF.
[10] Pendar, N. (2007). Toward spotting the pedophile telling victim from predator in text chats. In International Conference on Semantic Computing (ICSC 2007), Sep.
[11] Morris, C. and Hirst, G. (2012). Identifying sexual predators by SVM classification with lexical and behavioral features. In CLEF (Online Working Notes/Labs/Workshop).
[12] Parapar, J., Losada, D. E., and Barreiro, A. (2012). A learning-based approach for the identification of sexual predators in chat logs. In CLEF (Online Working Notes/Labs/Workshop).
[13] Peersman, C., Vaassen, F., Van Asch, V., and Daelemans, W. (2012). Conversation level constraints on pedophile detection in chat rooms. In CLEF (Online Working Notes/Labs/Workshop).
[14] Villatoro-Tello, E., Júarez-Gonźalez, A., Escalante, H. J., y Gómez, M. M., and Pineda, L. V. (2012). A two-step approach for effective detection of misbehaving users in chats. In CLEF (Online Working Notes/Labs/Workshop).
[15] Escalante, H. J., Villatoro-Tello, E., Júarez, A., Montes-y Gómez, M., and Villasẽnor, L. (2013). Sexual predator detection in chats with chained classifiers. In Proceedings of the 4th Workshop on Computational Approaches to Subjectivity, Sentiment and Social Media Analysis, June.
[16] Vartapetiance, A. and Gillam, L. (2014). "our little secret": pinpointing potential predators. Security Informatics, 3(1):3, Sep.
[17] Cheong, Y., Jensen, A. K., Guonadottir, E. R., Bae, B., and Togelius, J. (2015). Detecting predatory behaviour in-game chats. IEEE Transactions on Computational Intelligence and AI in Games, 7(3), Sep.
[18] Miah, M. W. R., Yearwood, J., and Kulkarni, S. (2011). Detection of child exploiting chats from a mixed chat dataset as a text classification task.

In Proceedings of the Australasian Language Technology Association Workshop 2011, December.

[19] Kontostathis, A., Garron, A., Reynolds, K., West, W., and Edwards, L. (2012). Identifying predators using chatcoder 2.0. In CLEF (Online Working Notes/Labs/Workshop).

[20] McGhee, I., Bayzick, J., Kontostathis, A., Edwards, L., McBride, A., and Jakubowski, E. (2011b). Learning to identify Internet sexual predation. International Journal of Electronic Commerce, 15(3):103–122.

[21] Pennebaker, J. W., Chung, C. K., Ireland, M. E., Gonzales, A. L., and Booth, R. J. (2011). The development and psychometric properties of liwc2007.

[22] Eriksson, G. and Karlgren, J. (2012). Features for modelling characteristics of conversations: Notebook for pan at clef 2012. In CLEF (Online Working Notes/Labs/Workshop).

[23] Morris, C. and Hirst, G. (2012). Identifying sexual predators by SVM classification with lexical and behavioral features. In CLEF (Online Working Notes/Labs/Workshop).

[24] O'Connell, R. L. (2003). A typology of child cyber exploitation and online grooming practices. Preston: University of Central Lancashire, Cybersex Research Unit.

[25] Olson, L. N., Daggs, J. L., Ellevold, B. L., and Rogers, T. K. K. (2007). Entrapping the innocent: Toward a theory of child sexual predators' luring communication. Communication Theory, 17(3).

[26] Morris, C. (2013, January 30). Identifying Online Sexual Predators by SVM Classification with Lexical and Behavioral Features (Master of Science Thesis). University of Toronto, Canada. Retrieved from tp.cs.toronto.edu/pub/gh/Morris,Colin-MSc-thesis-2013.pdf.

[27] Olson, L. N., Daggs, J. L., Ellevold, B. L., & Rogers, T. K. K. (2007). Entrapping the Innocent: Toward a Theory of Child Sexual Predators' Luring Communication. Communication Theory, 17(3), 231–251.

[28] Aggarwal, C. C. (2015). Mining Text Data. In *Data Mining* (pp. 429–455). Springer International Publishing. Retrieved from http://dx.doi.org/10.1007/978-3-319-14142-8_13.

[29] Duda, R. O., Hart, P. E., & Stork, D. G. (2012). *Pattern Classification* (second ed.). Wiley-InterScience.

[30] Ebrahimi, Mohammadreza. (2016). Automatic Identification of Online Predators in Chat Logs by Anomaly Detection and Deep Learning. 10.13140/RG.2.2.18105.01127.

4

CNN Classification Approach to Detecting Abusive Content in Text Messages

R. Dinesh Kumar[1], G. Vinoda Reddy[2], S. Ravi Chand[3], B. Karthika[4], and V. Murugesh[5]

[1]Professor – CSE Department, Siddhartha Institute of Technology and Science, Hyderabad, India.
[2]Professor of Computer Science and Engineering(AI & ML) CMR Technical Campus, Hyderabad, India
[3]Professor – ECE Department, Nalla Narasimha Reddy Group of Institutions, Hyderabad, India.
[4]Assistant professor – IT Department, PSNA College of Engineering and Technology,Dindigul, India.
[5]Assistant Professor – CSE Department, College of Informatics, Bule Hora University, Ethiopia
Emal: me.dineshkumar@gmail.com

Abstract

Currently, abusive messages or text are slowly eroding the effectiveness of the user experience on the Internet. As a result of new technology, the number of people prepared to abuse technology is rising. The major goal of this research work is to prevent abusive predators from engaging in abusive behaviour while also investigating various sorts of abusive patterns and messages. To address the issue, we use natural language processing (NLP) approaches to solve the challenge of recognizing abusive content in text sources. The machine learning techniques with three classifiers, support vector machines (SVMs), multilayer perceptron (MLP), convolutional neural network (CNN) and text encoder (Bag of Words), are applied to evaluate abusive text with better accuracy. The testing is carried out using a newly available dataset on the Reddit website. A CNN classifier and text encoders are integrated to produce a more efficient outcome based on performance metrics (precision, recall and F-measure).

4.1 Introduction

4.1.1 Humanity

Humanity is divided into two halves: male and female. Together, men and women may effectively run a society. Many of the horrors that exist are perpetrated against women, man has subjected women to his will, using her to promote his self-gratification, cater to his abusive pleasures, and be instrumental in advancing his comfort. However, he has never aspired to raise her to the position she was designed to fill. Even though women hold a unique position in society and the country, they have been subjected to the most heinous abuses without any justification or fault.

Women have been left behind as a result of the above. However, they have also been psychologically disempowered and notified of the hopelessness of their situation, making them feel physically and intellectually dwarfed. As a result, they have not only become targets of prejudice and exploitation, but they have also lost all motivation to defend themselves or catch up with the other half of the population.

Women's biological conditions make them easy prey, especially when it comes to physical dominance. As a result, women have become victims of their reproductive nature. The inherent distinction between men and women was viewed as a female flaw and women are referred to as the "weaker sex".

Teenagers make up almost a third of the population. With more youth in the country, the country can develop more efficiently in society, economically and educationally. People in this age group go through various physical and psychological changes, including abusive development, as shown in Figure 4.1. Teenagers engage in various abusive activities as a result of hormonal changes. In today's environment, digital communication is essential for interpersonal contact.

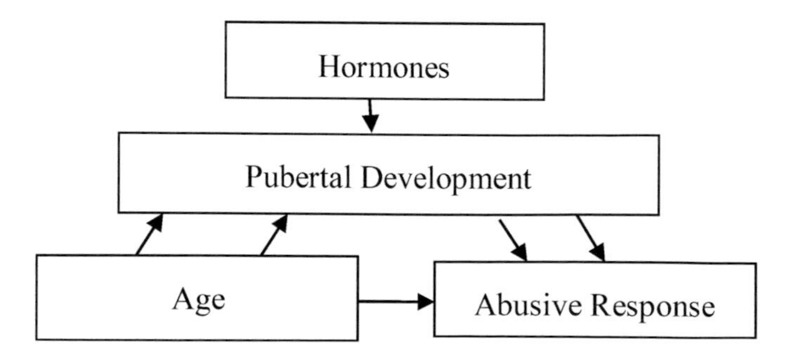

Figure 4.1 Scientific based physical development in youth.

4.1.2 Abusive harassment on the internet

Abusive harassment can take many forms; one is abusive cyber harassment, which is the subject of this work. Cyber abusive harassment is defined as harassment such as abusive messages, abusive images, and vulgar audio and erotic links with abusive content through different social media platforms.

Abusive harassment can have a variety of effects on an individual, including physical harm and emotional anguish and collapse.

4.1.3 Learning algorithm

In recent years, CNNs have implemented text, audio, language, word, and speaker recognition tests and have significantly improved the results. The CNN elements were unique and effective in improving the environment of abusive predator methods. Researchers in that discipline have done a limited number of studies and related work. According to the study's backdrop and review of the research problem, there has been a need for an effective abusive predator in a noisy environment.

The proposed method deals with the system by identifying patterns and following up on all text messages to prevent abusive predators.

4.2 Literature Survey

The proposed B-Spline curve recognizes vowels in palm leaf manuscripts in the Tamil language. It improves the results compared to the BFS algorithm. The proposed segmentation-based Omni font vocabulary uses a character segmentation algorithm and neural network and feature extraction. The approach yielded 95% accuracy at the font recognition stage. The future scope of this research includes generalization at the learning stage and reducing error rates using post-processing techniques. The future scope leads to the use of B-Spline curve recognition in addition to stroke recognition [1].

The various character recognition methods and literature review methods that are in existence are discussed, including artificial intelligence and machine learning. The comparison of CNN architecture with ML methods for character recognition was carried out in the Bangla language. The CNN architecture is built with different layers used in the Bangla datasets. The future work of this research is to analyse complex characters in the Bangla language [2], and researching various regional languages, analysing regional characters and inventing multilingual text recognition.

Pre-processing and recognition are proposed as a model where character recognition is better than previous algorithms. The proposed method with SVM yields better accuracy when compared to traditional methods. The future scope of the research work involves character recognition in real-world scenarios and using multi-lingual character recognition [3].

The proposed method can be applied to preventing crimes involving vehicles by scanning the vehicle number plate and using automatic number plate character recognition to monitor traffic and congestion control [4]. Firstly, the objectives are a large and complex handwritten character dataset (HMBD) and analysis using DL and CNN architectures with HMB1 and HMB2. The text dataset phase to recognition and the classification phase passing through segmentation and feature extraction phases are carried out. The future work of this work will help in preventing crimes involving vehicles.

It deals with different types of OCR available in the market and how the customer can use them, and discusses various south Indian languages, including Tamil, Telugu, Kannada, and Malayalam, for character recognition. This work also proposed character recognition based on neural networks, character normalization, correlation methods, and segmentation. The authors also concluded that CNNs perform better than other conventional methods for most languages. The future work of this research lies in applying OCR to websites and making them work effectively [5].

Comparative studies of different boosting algorithms have been made for recognition. The input dataset is taken from the mobile phones of different individuals. The findings have revealed that the proposed strategy has a high level of accuracy in classifying daily activities. These findings support the proposed system's ability to identify everyday living activities by using mobile data from the sensors and imply that it can help reduce daily immobility trends and aid positive living and pleasure [6].

Artificial intelligence uses different machine learning methods to predict machine failures. The proposed work predicts the failure in advance based on the present state of the equipment. It ensures excellent monitoring and prediction accuracy and a high level of confidence [7].

Investigations have been carried out in two ways: using the Hyper Opt framework to undertake systematic hyperparameter optimization (HPO). The accuracy scores and training times four datasets have been presented [8].

In changing lives with new technology, the main goal is to monitor abusive content in messages on social media platforms and generate alerts via an application. An algorithm was built to classify textual communication [9].

Currently, cyber harassment is a rising and severe issue. Both deep learning and machine learning algorithms have been applied in the existing text classification methods. The deep learning method yields a better result for text classification. MaLang, a revolutionary technique, has been created to detect abusive text content. MaLang outperforms the competition with a classification accuracy of 98.2% [10].

Cyber-abusive harassment commonly causes severe distress, especially among women and children. Many incidents of internet harassment have recently occurred around the world. As a result, experts are paying more attention to detecting abusive text messages on social media. This study aims to use natural language processing and machine learning to create an effective technique for detecting online abusive text messages [11].

This article builds on a prior study by describing a method for recognizing and preventing cyberbullying. It begins with a review of existing work in cyberbullying detection. Following that, a technique for identifying cyberbullying in Arabic content is classified and evaluated [12].

The new deep learning architecture is named "convolutional bi-directional LSTM (C-BiLSTM)" for recognizing improper query recommendations. C-BiLSTM outperforms in hand-crafted features and has a high precision rate [13].

All necessary references for detecting abusive language have been presented. They can be used as a resource for other academics looking for information about abusive language on Twitter relevant to their field of research [14].

The first step is to create a single categorization system, translate all writing into a universal language, and then classify them. The experimental results demonstrate that the proposed method obtains an F1-score and accuracy of over 93% and 91%, respectively [15].

4.3 Proposed Methodology

This study focuses on text classification or the process of categorizing text based on its content. This can be applied in various applications to classify the text. Classifying the abusive text in the content is a challenging task. The manual process is time-consuming. This research work aims to observe the usage of NLP and machine learning approaches that can aid in the detection of abusive text, as shown in Figure 4.2.

4.3.1 Pre-processing

Pre-processing is the process that helps to convert the given text into tokens. The text is converted into lowercase in pre-processing, and tokens are extracted from the phrases. All numerals and stop words are removed. The

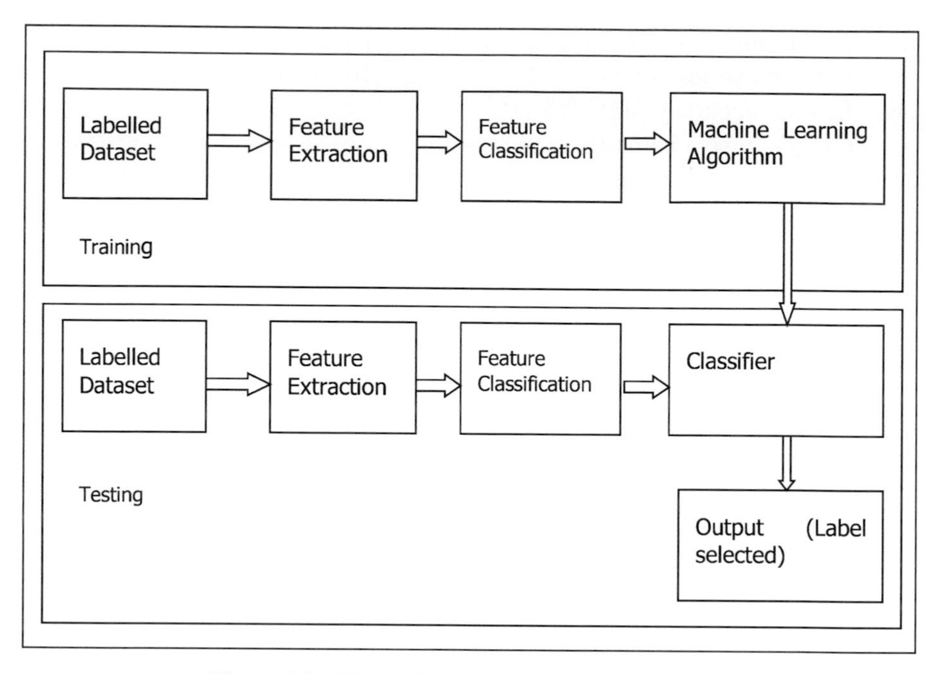

Figure 4.2 Block diagram for the proposed method.

next step is to extract various features from the pre-processed text and apply the classification process.

4.3.2 Feature extraction

The feature extractor is in charge of converting tokens into numerical values that a learning system can employ. The feature extraction method applied is Bag of Words (BoW).

4.3.3 Vector space model (VSM)

The VSM is the primary technique for handling the task of text encoding. The VSM method converts a text document into a numerical vector

4.3.3.1 Bag of words

One of the essential feature weighting strategies is the BoWs scheme. It also applies to other different applications like computer vision and document categorization. BoWs assigns a vector to each text that represents the probability for which a word from a dictionary exists in the text. A training collection of texts could be used to create that dictionary.

4.3.4 Classification methods

Classification is carried out by combining the extracted features from the preceding phases. The classification technique recognizes the character based on the rules that have been established. The class membership pattern is used in the classification process for decision making. The classification task is challenging. As a result, the feature extraction approach reduces the likelihood of misclassification. The classification procedure is completed after this, but classification becomes a problem when characters fall into an unknown pattern.

4.3.4.1 Support vector machine (SVM)

One of the binary and supervised learning classifiers is the support vector machine (SVM). The SVM classifier determines which category each belongs to by evaluating the subset of features. The hyperplane was first chosen to split the two classes to avoid data misclassification. The hyperplane was selected to partition the output class labels y {−1, 1} if the input text character feature is "x", and the character class outcome is y, and the hyperplane is specified as follows:

$$w \, . \, x + b = 0. \tag{4.1}$$

Then,

$$y_i \, (w \, . \, x + b) \geq 1 \text{ where } i = 1,2,3,\ldots N. \tag{4.2}$$

The hyperplane is computed using equations (4.1) and (4.2), and it is utilized to minimize the variance between the values in Figure 4.3.

The difference between the hyperplanes is then determined in the following equation

$$d_+ + d_- = \frac{2}{\|w\|}. \tag{4.3}$$

The newly generated features are compared with the existing feature space based on the data. Although this SVM-based classification has a high precision value, it has a low recall value. The threshold value is then utilized to fine-tune the SVM to improve recall and classification performance. As a result, the optimal hyperplane has been created, and the error function has been minimized using an iterative training approach.

4.3.4.2 Multilayer perceptron (MLP)

A multilayer perceptron (MLP) maps input data sets to a set of acceptable outputs. An MLP is a modified linear perceptron that can identify data. An

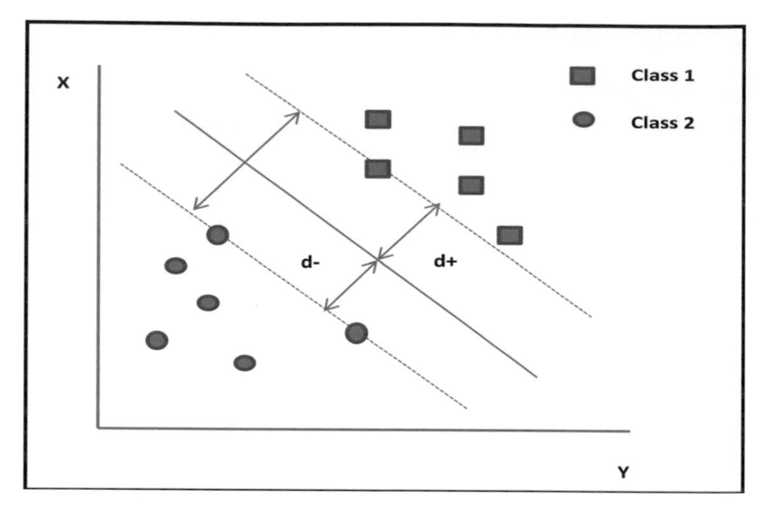

Figure 4.3 Support vector machine with hyperplane.

MLP is built with many layers of nodes in a directed graph, each one fully connected to the next is depicted in Figure 4.4. The non-linear activation function is applied to each neuron node except input nodes. An MLP trains the network via backpropagation, which is a supervised learning technique.

4.3.4.3 Convolutional neural networks (CNN)

Convolutional neural networks are used in image processing and data analytics. They are constructed by layers in which each neuron receives an input, computes a dot product, and optionally seeks non-linearity. The neurons in a CNN's layers are made in three dimensions: height, width, and depth. Convolutional-based layers make up most of a CNN is shown in Figure 4.5. These layers are a collection of input filters that extract a specific feature from the input, with the feature map serving as the targeted output. Every layer's filtered results are delivered through the feature space. The correlations are contained within the different filter representations. The multi-scale representation of the input text that recognizes the texture components incorporates the feature correlations of several layers.

The central unit of developing the CNN is the convolution layer, which is statistically controlled to merge a data set. Constructing the feature map is applied to the input data using the convolution filter. The operation is computed by moving the filter across the input data. A component-related matrix creation is made that forms the feature map at a specific point.

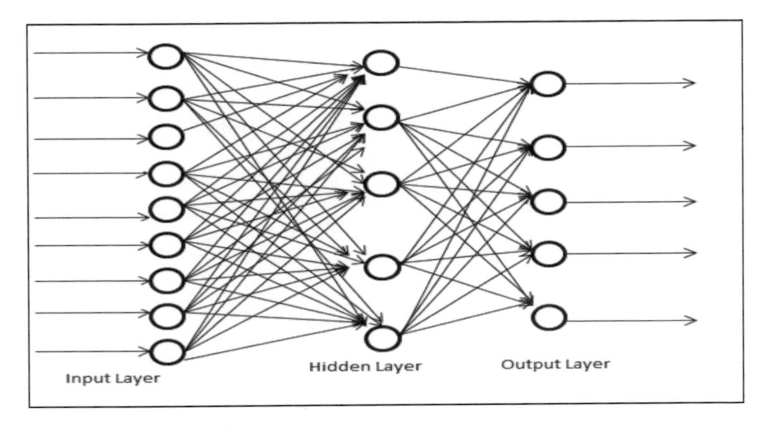

Figure 4.4 Multilayer perceptron.

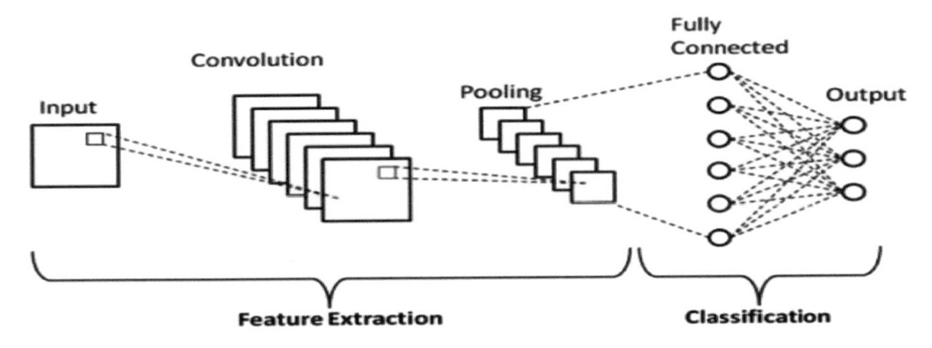

Figure 4.5 CNN architecture.

4.4 Performance Analysis and Metrics

The performance analysis of different classifiers is carried out for text classification to detect abusive messages. The following is a list of the examined performance metrics.

4.4.1 Precision

Precision, sometimes known as the positive anticipated value, is a metric for determining how closely retrieved characters are connected to user requests. The precision value is then determined as follows:

$$\text{Precision} = \frac{\left|\{\text{relevant document}\}\right| \cap \left|\{\text{Retrieved Document}\}\right|}{\left|\{\text{Retrieved Document}\}\right|}. \quad (4.4)$$

4.4.2 Recall

The parts of the document relevant to the user query that is successfully retrieved is referred to as recall. The following equation is used to calculate recall.

$$\text{Recall} = \frac{|\{\text{relevant document}\}| \cap |\{\text{Retrieved Document}\}|}{|\{\text{Relevant Document}\}|}. \tag{4.5}$$

4.4.3 F-measure

The F-measure is determined by adding the precision and recall values of each class.

4.4.4 Accuracy

Accuracy is calculated using the following equation (4.6): the number of positive texts divided by the total number of observed texts by the system.

$$\text{Accuracy} = \frac{\text{number of true positive} + \text{number of true negative}}{\text{number of true positive} + \text{false positive} + \text{false negative} + \text{true negative}}. \tag{4.6}$$

4.5 Results and Discussion

The performance analysis carried out with different classifiers, SVM, MLP and CNN, is presented in Table 4.1 with the nature of the message. Tables 4.2 and 4.3 define the accuracy and F-measure of different classifiers.

Figure 4.6 shows the comparison with the existing model [16]. We have highlighted the existing model regarding the accuracy and F-measure findings for clarity's sake.

Table 4.1 Comparison of precision and recall.

Classifier	Nature of message	Bag of Words	Precision	Recall
SVM	Abusive	BoW (2 Gram)	0.8356	0.8974
	Neutral	BoW (2 Gram)	0.8123	0.8642
MLP	Abusive	BoW (2 Gram)	0.8767	0.8839
	Neutral	BoW (2 Gram)	0.8835	0.8234
CNN	Abusive	BoW (2 Gram)	0.8962	0.9107
	Neutral	BoW (2 Gram)	0.9342	0.9621

Table 4.2 Comparison of accuracy.

Classifier	BoW
SVM	0.8019
MLP	0.7234
CNN	0.9352

Table 4.3 Comparison of F-Measure.

Classifier	BoW
SVM	0.7402
MLP	0.7213
CNN	0.9527

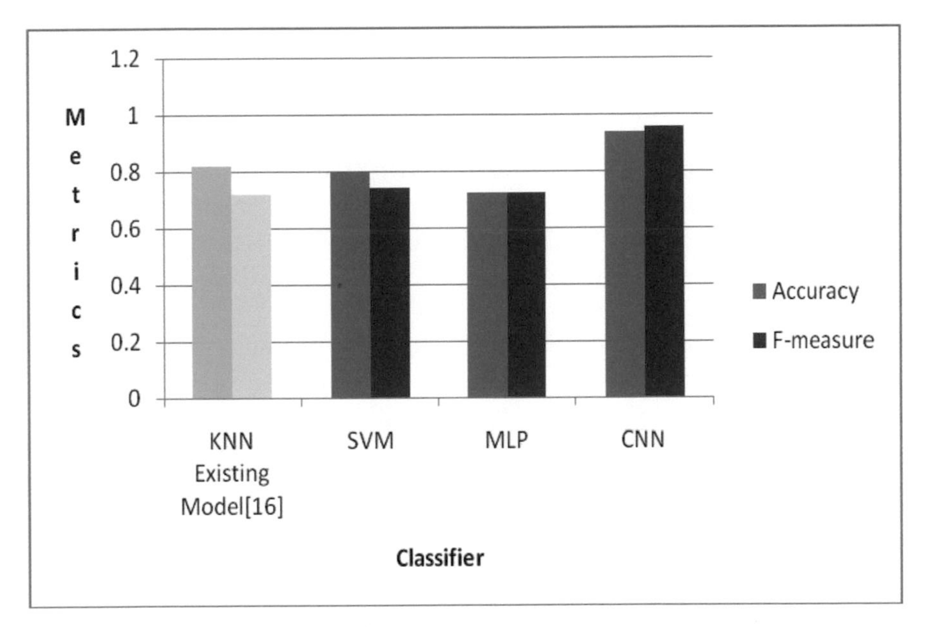

Figure 4.6 Comparison with the existing model of accuracy and F-measure.

4.6 Conclusion

The Internet in the real world has numerous dangerous areas where children may be exposed to unsuitable content such as abusive or disrespectful comments. We use natural language processing (NLP) approaches to solve the challenge of recognizing abusive content in text documents in this paper. Automated techniques are needed for the detection of abusive content provided by users. We use machine learning approaches to solve the challenge of detecting abusive text. The experiments were carried out using a dataset taken from the Internet based on publicly available information on the Reddit website. The outcome of the CNN classifier gives better results. The error obtained using the CNN classifier is minimized. The precision and recall of the proposed methodology yield higher performance and efficiency. These models can be applied to social networks and YouTube, where you can find many unregulated comments.

References

[1] S. Athisayamani et al., "Recognition of ancient Tamil palm leaf vowel characters in historical documents using B-spline curve recognition," Procedia Comput. Sci., vol. 171, pp. 2302-2309, 2020. doi:10.1016/j.procs.2020.04.249.

[2] A. Hazra et al., "Bangla-Meitei Mayek scripts handwritten character recognition using Convolutional Neural Network" Appl. Intell., vol. 51, no. 4, 2021, doi:10.1007/s10489-020-01901-2.

[3] Y. B. Hamdan and Sathish, "Construction of Statistical SVM based Recognition Model for Handwritten Character Recognition," J. Inf. Technol. Digit. World, vol. 3, no. 2, pp. 92-107, 2021. doi:10.36548/jitdw.2021.2.003.

[4] R. Dineshkumar and J. Suganthi, "Sanskrit character recognition system using neural network," Indian J. Sci. Technol., vol. 8, no. 1, p. 65, 2015. doi:10.17485/ijst/2015/v8i1/52878.

[5] S. Chandra et al., "Optical character recognition – A review," Int. Res. J. Eng. Technol. (IRJET), vol. 7, no. 4, 2020.

[6] S. Rahman et al., "Performance analysis of boosting classifiers in recognizing Activities of Daily Living," Int. J. Environ. Res. Public Health, vol. 17, no. 3, p. 1082, 2020. doi:10.3390/ijerph17031082.

[7] G. M. Tina et al., "A state-of-art-review on machine-learning based methods for PV," Appl. Sci., vol. 11, no. 16, p. 7550, 2021. doi:10.3390/app11167550.

[8] S. R, S. S. Ayachit et al., "Competitive analysis of the top gradient boosting machine learning algorithms," 2nd Intl. Conf. on Adv. in Compute. Communication Control and Networking (ICACCCN), vol.5 2020, pp. 191-196. doi:10.1109/ICACCCN51052.2020.9362840.

[9] R. Gandhi, "Termination of cyber-abusive harassment and abuse with teenagers using artificial intelligence," Educ. Quest An Int. J. Educ. Appl. Soc. Sci., vol. 11, no. 3, pp. 169-174, 2020. doi:10.30954/2230-7311.3.2020.3.

[10] P. Kompally et al., "MaLang: A decentralized deep learning approach for detecting abusive textual content," Appl. Sci., vol. 11, no. 18, p. 8701, 2021. doi:10.3390/app11188701.

[11] M.Islametal.,"CyberbullyingDetectiononSocialNetworksUsingMachine Learning Approaches", (2021). 10.1109/CSDE50874.2020.9411601.

[12] B. Haidar et al., “A multilingual system for Cyberbullying detection: Arabic content detection using machine learning” [Journal], Adv. sci. technol. eng. Syst. j., vol. 2, no. 6, pp. 275–284, 2017, doi:10.25046/aj020634.

[13] Yenala et al., “Deep learning for detecting inappropriate content in-text”, *International Journal of Data Science and Analytics*, Vol. 6, (2018), doi: 10.1007/s41060-017-0088-4.

[14] Naseem et al., “Abusive Language Detection: A Comprehensive Review”, *Indian Journal of Science and Technology*, Vol. 12, Pp.1–13, (2019) doi:10.17485/ijst/2019/v12i45/146538.

[15] Fatima-zahra El-Alami et al.,”A multilingual offensive language detection method based on transfer learning from transformer fine-tuning model”, *Journal of King Saud University - Computer and Information Sciences*, 2021, doi:10.1016/j.jksuci.2021.07.013.

[16] Gonzalo Molpeceres Barrientos et al.,” Machine Learning Techniques for the Detection of Inappropriate Erotic Content in Text”, *International Journal of Computational Intelligence Systems,* Vol. 2(1), 2020, doi /10.2991/ijcis.d.200519.003.

5

Detection of Online Sexual Predatory Chats Using Deep Learning

R. Kesavamoorthy[1], S. P. Anandaraj[2], T. R. Mahesh[3], V. Rajesh Kumar[4], and Asadi Srinivasulu[5]

[1]Associate Professor, Department of Computer Science and Engineering, CMR Institute of Technology, Bengaluru
[2]Associate Professor, Department of Computer Science and Engineering, School of Engineering, Presidency University, Bangalore
[3]Associate Professor, Department of Computer Science and Engineering, Faculty of Engineering and Technology, Jain Deemed-to-be University, Bangalore
[4]Assistant Professor, Department of Electrical and Electronics Engineering, Sir M Visvesvaraya Institute of Technology, Bengaluru, India.
[5]Data Science Research Lab, BluCrest University- 91016, Monrovia, Liberia
E-Mail: kesavamoorthycse@gmail.com; anandsofttech@gmail.com; maheshtr.1978@gmail.com; vrk5197148@gmail.com; head.research@bluecrestcollege.com

Abstract

The availability of various online messaging services enables anyone to interact with everyone. In this scenario, children or even teenagers are falling prey to online predators who pose as innocent users and try to manipulate children's emotions for sexual purposes. This is how child exploitation happens online. Therefore, it is necessary to analyze the prey and predators' conversations and classify them as good or bad. Various approaches have been in practice for the last two decades, and the latest one uses the concept of deep learning. Having its root in machine learning and neural networks, deep learning is a concept that has evolved using the structure and function of the human brain as inspiration. Although initially deep learning was used only for classification problems, now it is being used for solving problems in various domains. This chapter will first introduce various machine learning models to identify online sexual predators, and then it will present the various deep learning models to solve the same problem. The last part of this chapter will compare the performance of deep learning models over machine learning models. This chapter may be considered as a journey towards the problem of identifying online predators and solving the problem using deep learning.

5.1 Introduction

As well as disasters, crimes, and other threats, public safety should also consider providing a safe environment for children on social media. In general, grooming is the process of a predator taking advantage of the prey's emotions to satisfy their sexual intentions physically. If the same happens online, it is cyber grooming. Analyzing offline conversations to identify chat predators is only post mortem. Therefore, automated methods are required to identify the predators and the predatory chats during the conversations. Considering this objective, Figure 5.1 shows the explored four hypotheses.

5.2 Machine Learning Models to Detect Online Sexual Predatory Chats

Child exploitation has increased considerably over the last few years, both physically and online, even after 17 years of the UN Convention on the Rights of the Child. Figure 5.1 shows the various AI tools that are available as of now to fight child abuse.

These tools focus on identifying child abuse online by using documents, images, web traffic and advertisements. However, the problem is identifying

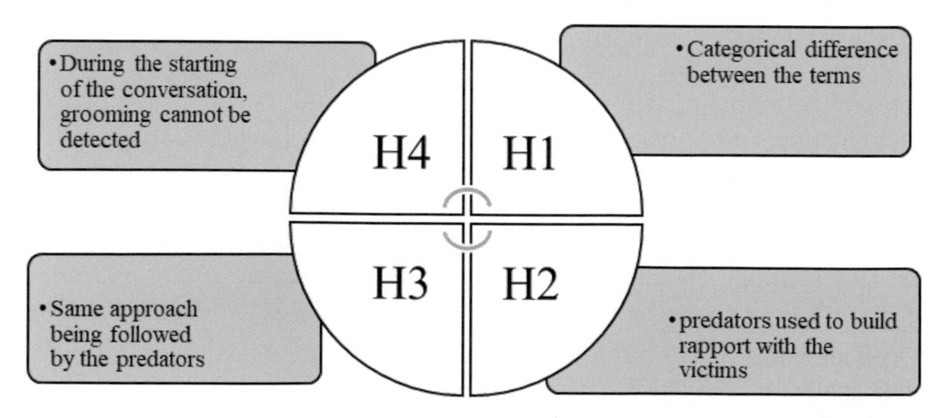

Figure 5.1 Hypotheses for the problem.

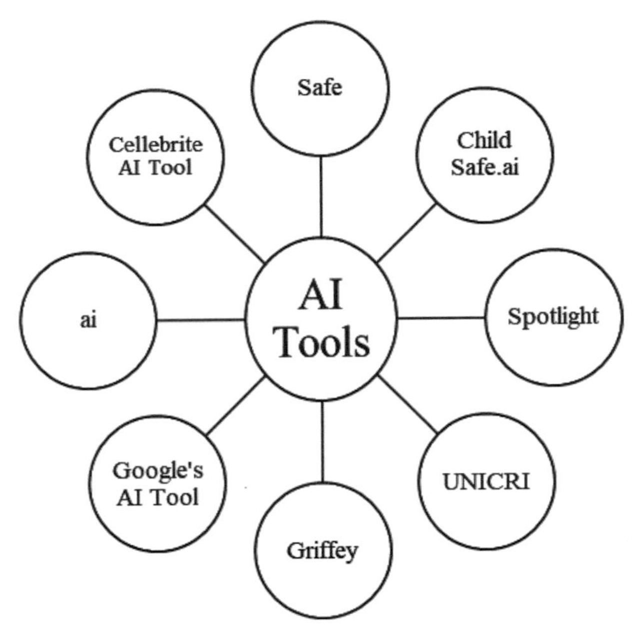

Figure 5.2 AI tools to detect online predators.

predatory conversations by using the chat or text messages between the predator and the victim. Machine learning and deep learning are most widely used to solve problems of this kind. Figure 5.3 shows the various machine learning models used to solve this problem. The following are the various models using deep learning to solve the problem:

a. Recursive neural network
b. Recurrent neural networks
c. Long short-term memory and
d. Convolutional neural networks.

This chapter show how deep learning methods can solve the problem of identifying online sexual predatory chats and how those methods outperformed when compared to machine learning models in solving the same problem.

5.2.1 Deep learning

Deep learning is a recent trend among researchers regarding data collection, data analysis, and data interpretation. Having its base in artificial neural networks, deep learning comes under the family of machine learning. The term

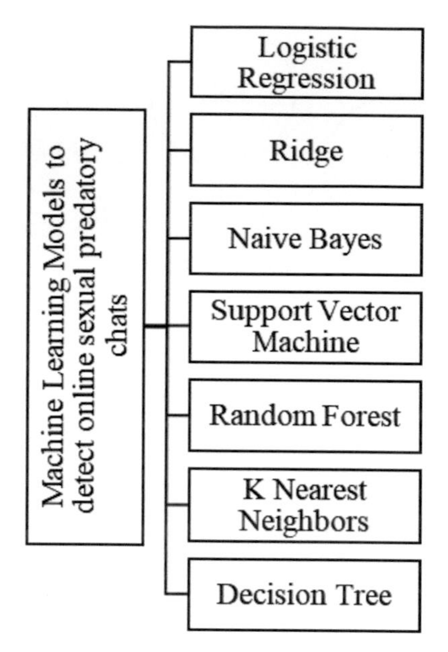

Figure 5.3 Machine learning models to detect online sexual predatory chats.

deep has its source in the logic of data passing through multiple hierarchies of layers. Each algorithm transforms the data in a non-linear fashion towards the output. Deep learning automates the process of predictive analytics, thereby making it easier and faster than machine learning models.

Deep learning models can be architected in deep neural networks, deep reinforcement learning, convolutional neural networks, recurrent neural networks, and deep belief networks. Every architecture has its own set of applications in virtual assistants, healthcare, robotics, entertainment, fake news detection, etc.

In particular, concerning the problem of predator identification on online chat, the following architecture is used, and the same is pictorially represented in Figure 5.4:

- Recursive neural network
- Recurrent neural networks
- Long short-term memory and
- Convolutional neural networks.

5.2.1.1 Recursive neural network

A new recursive neural network architecture was introduced in [1]. Identification of the recursive structure was the initial task. It helps identify

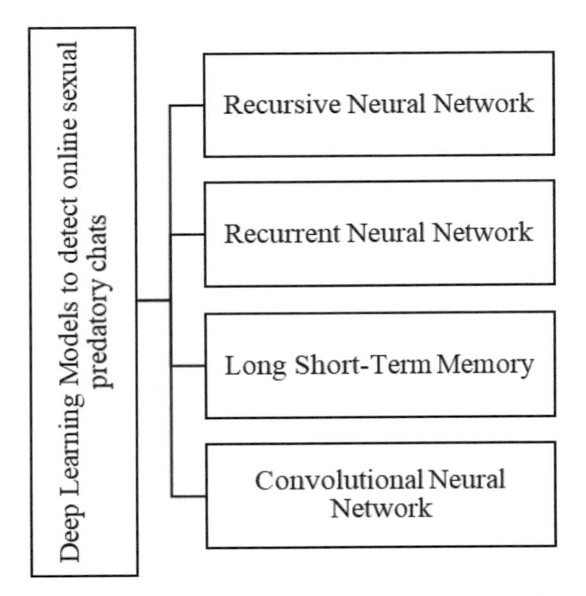

Figure 5.4 Deep learning models to detect online sexual predatory chats.

the segments and their interaction of forming as a whole. In addition to merging the image segments, this method also merges natural language words as a syntactic parser. The heart of this model is the learned semantic transformation of the original features. This approach works for segmentation, annotation and scene classification. The same was tested with the Penn Treebank and was claimed to have outperformed the previous approaches by 78%.

Expressing longer phrases was still a challenge for semantic word space. While moving towards compositionality, supervised training, powerful models, and evaluation resources are required for sentiment deduction. A sentiment treebank was introduced in [2] to overcome this problem. That treebank has more than 200 thousand phrases in the parse trees of over 11 thousand sentences. This became another threat, and it was solved by introducing recursive neural tensor networks, known as RNTN. This model has increased the performance by 5.4% more than the previous approach, even after incorporating the new challenges and new evaluation metrics.

5.2.1.2 Recurrent neural networks

Based on the principle of ANN, the recurrent neural network uses sequential data in which the output from the completed step will be passed as input to the current step. This makes it different from the traditional neural network in which the inputs and outputs are independent of each other. The heart of the RNN is the hidden state that is used to remember information about a sequence. The RNN can also process time-series data, natural language processing, image captioning, etc.

In the case of a bag of words approach wherein the model considers each word independently, the RNN is used to process sequential input data of variable size in which subsequent inputs are affected by the observed input. This makes RNN efficient by simply considering that the output at time "t" depends on the output at "$t-1$"[3].

Unlike other neural networks, the RNN reduces the complexity of parameters by having a memory that saves all the calculated values. This property makes it more efficient for processing input sentences. Problems of the type of opinion mining in sequential analysis can also be approached through the RNN [4]. This was treated as a kind of conditional random field and was claimed to outperform the existing methods. The most important thing to consider is that the claimed performance was achieved without utilizing syntactical analysis and sentiment lexicon.

5.2.1.3 Long short-term memory

Traditional backpropagation algorithms consider the partial derivatives in every preceding time stamp. A similar variant of the learning algorithm known as back propagation through time is used by the RNN. It is required to calculate more gradients because of too many time stamps. This problem was overcome by the introduction of long short-term memory [5].

When using deep learning for natural language, word embedding is one of the essential ideas concerning representing words. Before this, the traditional bag of words approach was used to analyze the document by using feature vectors such as the frequency of each word. By contrast, LSTM is used to represent each word in its vector. Later, at the end, the feature matrix is formed by concatenating the vector representations. This individual vector representation is known as embedding. This technique was first introduced in [6] based on backpropagation errors.

Over time, the techniques based on word embedding are called language models. One such significant model proposed by [7] has outperformed all other models. Mainly, the method used is known as the skip-gram method and is now popular with researchers in the name of word2vec and is used to represent sentences and paragraphs.

5.2.1.4 Convolutional neural networks

The convolutional neural network is one of the deep learning algorithms under the class of artificial neural networks. It is shortly known as CNN or ConvNet and also known as SIANN, i.e. shift invariant ANN. Although it was initially used only for image inputs, it is now being used widely for audio or speech inputs. Promising results have been obtained in the field of NLP and text mining. CNN exhibits its superiority over other NNs by reducing the required pre-processing on various datasets. CNN differentiates itself from other neural networks by having three layers such as

- the convolutional layer
- the pooling layer
- the fully connected layer .

How CNN's can be designed in a step-by-step way for sentence classification was first given in [8][9]. The authors portrayed a sentence classification with just one layer for convolution and pooling. This is common for almost all text classification, and to help readers for a better understanding, it is described here. It is even called a holistic bible in CNN history.

The word embeddins in a concatenated form will be the input that will be split into several regions so that each region has its filters. The filters here are nothing but the features after linear transformation. Then the input sequencc will be passed through the filters to form a vector. The actual convolution happens as the filter removes a few connections leading to a partial network. However, the actual problem lies in the output, as it will vary in size.

Moreover, this is when a pooling layer is required to make it a single value by an average of all the elements. This is why the pooling layer is placed after the convolution layer. There is another reason too. Pooling helps the CNN to imprison the characteristics of the abstract data. This work done by the pooling layer is called a subsampling procedure. As a result of this subsampling procedure, more and more abstract features will be captured at each layer. Therefore, the output layer will become a fully connected layer. The output values will be of fixed size [10].

A unified language framework is proposed by [11]. A specific word embedding approach is used for various NLP tasks such as chunking, semantic role labelling, and tagging part of speech. With two approaches in this architecture such as window and sentence, the main logic will be like the encoding of the first layer and will be done by using the index in the dictionary. In the case of the window approach, the number of words is fixed, and in the case of the sentence approach, the number of words will vary depending on the size of the sentence.

The authors of [12] used word embedding, specifically word2vec, to get the feature vectors and then applied the actual CNN to work on sentiment analysis and sentence classification. With only one layer for convolution and pooling, the method produced surprising results using a SoftMax classifier.

By enhancing the pooling mechanism to obtain semantic relations and form feature graphs, dynamic CNNs were used in [11] for sentence classification. Because of the complications in the model, it was not used widely.

One of the milestones of online predator identification, proposed in [13], used a deep convolutional neural network. The three significant challenges that researchers face while architecting a model for detecting predatory conversations are:

1. Although the RNN works well for a small set of words or sentences, it becomes intractable for input sequences with significant phrases, paragraphs, and many sentences.
2. Word embedding techniques like word2vec work well on general problems but not so well for online predator identification because it obtains its training data from news datasets and web documents that will lead to misclassification.
3. The space for the feature has become highly sparse with the bag of words approach.

To address these challenges, in [14] the feature vectors were directly fed into the convolution layer to learn them internally.

Dataset	Set	Characteristic	Value
PAN2012 Dataset	Training set	• Sample size	• 66,927
		• No. of predatory conservations	• 2016
		• No. of Unique users	• 97,695
		• No. of Unique predators	• 3737
	Testing set	• Sample size	• 155,128
		• No. of predatory conservations	• 3737
		• No. of Unique users	• 218,716
		• No. of Unique predators	• 254

Figure 5.5 Characteristics of PAN 2012 dataset.

This particular approach was processed in K80 GPU 2496 processor cores equipped with 24 GB of physical memory by taking advantage of the hardware level parallelism and neural network-based learning models. Thanks to Perverted-Justice, who built the PAN2012 dataset by posing as juveniles. This dataset is made public now and is widely used by almost all the solutions for online predator identification problems. Figure 5.5 shows the characteristics of the PAN2012 dataset.

Figure 5.6 shows a comparison of deep learning tools and their pros and cons.

The dataset is arranged into two XML files, one for training and another for testing. The XML files are parsed to extract the messages of each participant from each conversation. A sample of this is shown in Figure 5.7.

The performance of this method has been compared by setting up three groups of experiments such as:

1. ML models
2. CNN models with word vector
3. CNN models without word vector explicitly.

Furthermore, Figure 5.8 shows the group-wise results. It can be noted that the underlined one represents the best performance in each group.

Framework	Language	Algorithm Coverage	Developer	Characteristics
Caffe	C++/Cuda	CNNs	Berkley Vision and Learning Center	• Highly efficient for image processing with ConvNets • Specific to image and machine vision
Theano/ PyLearn	Python	• Restricted Boltzmann Machines • Stacked Denoising Autoencoders • CNNS	LISA lab at the University of Montreal	• General-Purpose • Requires symbolic math expressions
Torch	Lua Script	• Restricted Boltzmann Machines • Stacked Denoising Autoencoders • CNNS	Facebook and Google	• Matlab-like script and Intuitive in usage • General-Purpose • Learning curve for Lua language
ConText	C+/Cuda	• Supervised CNN • Semi-supervised CNN	Johnson	• High Performance • Specific to document classification • Easy-to-use bash script support • Runs on Nvidia GPU
DL4J	Java and Scala	• Restricted Boltzman Machines • CNNS • Recursive Nets • Recurrent Nets • Deep-belief Nets • Stacked Denoising Autoencoders • Deep Autoencoders	Sky Mind Company	• Faster than Python • General-purpose • Transparent parallelism • Work with Hadoop and Spark • Slower development speed compared to scripting languages
CNKT	C++/Cuda	• CNNs • Recurrent Nets • Long Short term Memory Networks(LSTMs)	Microsoft	• Graphical User Interface • Runs on multiple GPUs on multiple machines
Tensor Flow	C++/Python	• Multi-Purpose	Google	• Multi-platform (CPU, GPU and Mobile Device) • General purpose

Figure 5.6 Comparison of deep learning tools.

```
<conversation id =”8ff4c51529c81dabb0978206cb6bf06a”>
...
<message line=”4”>
        <author>f4113d73c0b80c35c5e01f76ab4</author>
        <time>12:34</time>
        <text>udidn’t talk 2 me yesterday</text>
</message>
<message line=’5’>
        <author>47243a4a2c68f2f00899670d455a21fa</author>
        <time>12:34</time>
        <text>I wasn’t on </text>
</message>
        <message line=”6”>
        <author>47243a4a2c68f2f00899670d455a21fa</author>
        <time>12:34</time>
        <text>Sorwy</text>
</message>
<message line=”7”>
        <author>4723a4a2c68f2f00899670d455a21fa</author>
        <time>12:35</time>
        <text>I got ur msg thoe..</text>
</message>
<message line=”8”>
        <author>f4113d73c0b80c35c5e01f736ab4</author>
        <time>12:36</time>
        <text>what are you doing?</text>
</message>
<message line=”9”>
        <author>47243a4a2c68f2f00899670d455a21fa</author>
        <time>12:37</time>
        <text>Workn..</text>
</message>
...
</conversation>
```


Figure 5.7 Sample snippet of a conversation.

5.3 Conclusion

This chapter focuses on introducing the problem of identifying the online predatory chats first and then the various solutions available to overcome them. It is very clearly illustrated that the deep learning algorithms outperformed the machine learning models in many ways. Particularly in deep learning, the deep convolutional neural network provides promising results. In addition to the performance of the algorithms, massive hardware support concerning parallelism is also required to achieve better results. RNN and LSTM, although they suffer from limitations, are found to be efficient for speech recognition and text analysis. As text analysis is the heart of solving

Learning scheme	Exp. No.	Settings	Precision (%)	Recall (%)	F1 - Score (%)
SVM 1	1	linear kernel	78.13	50.06	61.02
NN	2	nodes: 2000, encoding: frequency of unigram, vocab. size ¼ 5000	91.71	70.71	79.85
	3	nodes: 2000, encoding: frequency of bigram, vocab. size ¼ 7000	92.14	68.54	78.60
Pre-trained W2V-CNN	4	nodes:2000, skip-gram model with negative sampling, pooling type: max, word vector dimension: 3M x 100	86.41	70.94	77.91
W2V-CNN	5	skip-gram model with hierarchical softmax, nodes:2000, pooling type: max, word vector dimension:36,314 x 100	89.24	70.71	78.90
	6	skip-gram model with negative sampling, nodes:2000, pooling type: max, word vector dimension:36,314 X 100	89.46	73.30	80.58
GloVe-CNN	7	nodes:2000, pooling type: max, word vector dimension:36,350 x 100	91.02	72.22	80.54
BOW-CNN	8	nodes:2000, region size:(8 and 15), encoding: binary-encoded unigram, vocab. size = 5000, pooling type: max	92.52	70.94	80.31
One-hot CNN	9	nodes:2000, region size:(1,2 and 3), encoding: concatenation of one-hot vectors, vocab. size = 5000, pooling type: max	91.57	72.41	80.87

Figure 5.8 Group-wise performance comparison.

the problem of predatory chat identification, DCNN can further be extended with supervised DCNN.

5.4 Acknowledgements

We would like to record our sincere thanks to our institutions, CMR Institute of Technology, Bengaluru, Presidency University, Bangalore, Jain Deemed-to-be University, Bangalore, and Shridevi Institute of Technology, Tumkur, wherein we obtained access to the various resources to prepare this book chapter.

References

[1] Socher R, Lin CC-Y, Ng AY, Manning CD. Parsing natural scenes and natural language with recursive neural networks. In: Getoor L, Scheffer T, editors. ICML. Omnipress; 2011. p. 129e36.

[2] Socher R, Perelygin A, Wu J, Chuang J, Manning CD, Ng A, et al. Recursive deep models for semantic compositionality over a sentiment treebank. In: Proceedings of the 2013 conference on empirical methods in natural language processing. Seattle, Washington, USA: Association for Computational Linguistics; 2013. p. 1631e42.

[3] Mikolov T, Karafiat M, Burget L, Cernocký J, Khudanpur S. Recurrent neural network based language model. In: Interspeech 2010, 11th

annual conference of the international speech communication association, Makuhari, Chiba, Japan, September 26e30, 2010; 2010. p. 1045e8.
[4] Irsoy O, Cardie C. Opinion mining with deep recurrent neural networks. In: Proceedings of the conference on empirical methods in natural language processing; 2014. p. 720e8.
[5] Goodfellow I, Yoshua B, Courville A. Deep learning. 2016. Book in preparation for MIT Press.
[6] Rumelhart DE, Hinton GE, Williams RJ. Learning representations by backpropagating errors. Nature 1986;323(6088):533e6.
[7] Mikolov T, Sutskever I, Chen K, Corrado GS, Dean J. Distributed representations of words and phrases and their compositionality. In: Presented at the advances in neural information processing systems 26: 27th annual conference.
[8] Zhang Y, Wallace B. A sensitivity analysis of (and practitioners' guide to) convolutional neural networks for sentence classification. CoRR, abs/1510.03820. Retrieved from. 2015. http://arxiv.org/abs/1510.03820.
[9] Anthoniraj, S., P. Karthikeyan, and V. Vivek. "Weed Detection Model Using the Generative Adversarial Network and Deep Convolutional Neural Network." *Journal of Mobile Multimedia* (2021): 275–292.
[10] Deepajothi, S., D. Palanival Rajan, P. Karthikeyan, and S. Velliangiri. "Intelligent Traffic Management for Emergency Vehicles using Convolutional Neural Network." In *2021 7th International Conference on Advanced Computing and Communication Systems (ICACCS)*, vol. 1, pp. 853–857. IEEE, 2021.
[11] Collobert R,Weston J, Bottou L, Karlen M, Kavukcuoglu K, Kuksa PP. Natural language processing (almost) from scratch. JMLR 2011;12:2493e537.
[12] Kim, Y. (2014). Convolutional Neural Networks for Sentence Classification. In Proceedings of the Conference on Empirical Methods in Natural Language Processing. Doha, Qatar.
[13] Kalchbrenner N, Grefenstette E, Blunsom P. A convolutional neural network for modelling sentences. ACL 2014;2014.
[14] Ebrahimi, Mohammadreza. (2016). Automatic Identification of Online Predators in Chat Logs by Anomaly Detection and Deep Learning. 10.13140/RG.2.2.18105.01127.

6

Enhancing ATM Security in the Forensic Domain Using Artificial Intelligence

M. S. Swetha[1], M. S. Muneshwara[2], Ashutosh Raj[3], Atul Tomar[4], Ayush Prakash[5], and Chetan Singh[6]

Assistant Professor, Dept. of ISE, BMS Institute of Technology and Management, Avalahalli, Yelahanka, Bangalore - 64
Email: swethams_ise2014@bmsit.in; muneshwarams@bmsit.in; 1by19is031@bmsit.in; 1by19is032@bmsit.in; 1by19is033@bmsit.in; 1by19is044@bmsit.in

Abstract

Almost every day, we hear news related to various frauds and scams relating to ATMs. ATMs came as a replacement for the traditional banking system. However, the lack of a proper security system has opened up new threats. Banks, however, seeing this increase in fraud related to ATM transactions, came up with one-time-password based transactions. Although this may help prevent unauthorized transactions to some extent, there is still scope for improvement in avoiding false transactions. This chapter focuses on securing transactions using facial recognition. Facial recognition is one of the most secure methods for security as it is fast and reliable. According to our experiment, we found that this can help enhance security by many times. The scope of error is very minimal and is accurate most of the time. Our experimental outcome showed accuracy of up to 95% with various angles of faces tested. We tried to utilize unique features provided by facial recognition at ATMs. When a user swipes their card, they undergo a facial scan. After scanning, the face undergoes processing through OpenCV, free for use software under the open-source BSD license, to process the scanned face. After processing, the request is made to a database to fetch the cardholder details (details like the card owner's face) and then matched with the scanned face. If the scanned face matches, the user can complete their transaction, thus minimizing fraud and scams. Facial recognition technology helps the machine identify every user uniquely, thus making a face and a card number a "key" in the process. This completely eliminates the chances of fraud due to theft and duplication of ATM cards. This chapter also focuses on linking a single card with multiple users if they belong to the same family. All the family members can use the same ATM card after adding their facial features to the database.

6.1 Introduction

Digital forensics is a field of forensic science that deals with the recovery, investigation, examination, and analysis of data found in digital devices that are frequently used in mobile devices and computer crime [1].

The days are gone that one needs to visit a bank to withdraw cash thanks to ATM. ATM stands for automatic teller machine. An ATM is an electronic telecommunication device that allows customers to manage their financial accounts, view statements, and withdraw cash [2-3]. An ATM is a self-service terminal that uses a secure method of encrypted communication through the internet to provide these facilities to the customer. For ATM service, every customer has an individual plastic card issued by a bank containing a magnetic strip called an ATM card, used to authenticate them and operate ATMs. ATMs swipe cards in a slot that reads the strip instantly and returns them after the customer has done operating their account [4-6]. So, the first step to operating an ATM is inserting the card, followed by the ATM asking the user to authenticate themselves by entering a PIN. The PIN, which stands for personal identification number, is a unique 4–6-digit long number and contains sensitive information and hence is kept secret [8]. In the case of entering the incorrect pin three times, the ATM card is blocked to prevent misuse and can only be unblocked once a customer visits the bank. After successfully validating the pin, the customer is authenticated and displayed with various menus [9–11].

The main problem with this kind of system is a lack of security. There is a single level of security that can be bypassed by spying, skimming, stealing, etc. One of the other problems is the current Covid-19 pandemic, where customers need to press keys to give various input physically. Withdrawing with a smartphone is one technique to overcome this problem. Customers can swipe their card, and the ATM will generate a temporary URL or a temporary QR code (which will redirect to the URL on scanning) which can only be accessed if a smartphone is connected to the ATM's Wi-Fi. Opening the URL will bring you to an ATM interface on your smartphone. This will significantly reduce the number of times a customer touches the ATM.

In this chapter, we aim to introduce adequate two-level security that can be used to enhance ATM security to a significant level and reduce the need for customers to touch the keypads [13-14]. A digital camera employed for facial recognition algorithm will be the digital device used in this experiment. The machine stores every image that the system reads in our database. We will use an alert mechanism to prioritize any records that fail the facial recognition algorithm for the third time.

Having an extensive database ensures that the scanned face may be matched with an existing record and that all relevant facts can be taken at any time, making the individual committing a crime extremely vulnerable. This strategy aids us in the investigation phase in the event of digital theft. Because we already have digital evidence, it will be easier for the appropriate authorities to investigate the theft. The user will be notified right away. If the user fails for the third attempt, a feature can be included to warn a security guard or play a loud sound [16].

6.2 Literature Survey

After a thorough search and evaluation, the literature review of the documents that support this system is represented below.

Sayan Hazra et al. concluded that one of the most important services is the ATM service. Around 3.5 million ATMs are placed worldwide, according to the ATM industry organization (ATMIA). Because no ATM card is necessary for transactions in this proposed model, facial recognition, aliveness check of the face, fingerprint verification, and OTP (one-time-password) verification are all recommended. Facial recognition is accomplished in this system by cascade classifiers, including Haar feature extraction. Different square-shaped forms are tracked to analyze the frame across the images. The paper proposes a system for a cardless and fast money withdrawal service, in which security is provided by facial recognition and aliveness checks, accompanied by fingerprint checks. Secondary user transactions are also possible using the account holder's stored information, where OTP verification has increased security by many times. The working model's demonstration result validates the suggested smart, secure, and rapid ATM service idea [1].

Egho-Promise et al. used Wireshark and packet snipper tools to analyse the ATM traffic data and determine who had stolen money from which account. This approach is beneficial for digital forensics officers to create the evidence against the criminal, but using this approach is quite complex [2].

Priyanaka Hemant kale et al. concluded that crime has been increasing rapidly with the increase in daily usage of ATMs for the past three years from 2016 to 2018. Moreover, one of the various tactics used by fraudsters is inserting malware into ATMs, which ultimately results in leakage of user's sensitive/crucial information which might be used to harm the user in numerous ways such as manipulating the machine to give out every cent to the fraudster. There are other infamous techniques such as card skimming and card trapping. In card skimming, the skimmer devices look precisely like a cardboard sheet slot that has been always present in ATM making it harder to discover, resulting in the ATM being unsafe to use. Moreover, in card trapping, the fraudster cleverly installs tools that disable the cash dispenser, which stops the users from picking up the cash withdrawn from the ATM. There are similar methods such as shoulder surfing. The fraudster installs a fake input taking device on top of the original ATM input device, making it seem like everything is all right. When the user enters their PIN, the fake module records the PIN, making the user's account vulnerable.

In order to stop these crimes, the proposed system uses a microcontroller that does not depend on the processor and a GSM can also be installed, which basically stands for global system for mobile communication and is just a digital cellular system for cell phones, it communicates serially with devices like a microcontroller. In order to make it more secure, one can install a fingerprint sensor, keypads and LCD to verify and guide the users during the usage of ATMs [3].

Srivatsan Sridharan et al., describes how an ATM is a machine that provides its user with instant money. However, before the user can withdraw the amount, they have to give a secure PIN as an input to the machine. The primary focus behind inventing this system was to provide cash to the bank's customers at lightning speed when needed. In the existing machines, there is a lack of security for assurance other than the secrecy of the PIN [4]. Table 6.1 shows the comparison different ATM security model.

6.3 Problem Statement

Nowadays, there is an increase in ATM fraud cases like card shimming, card skimming, card trapping, etc. The ATMs are not well equipped to detect this kind of fraud. Moreover, it is impossible to tell which person is withdrawing money from the account. So, in order to tackle these problems, this chapter introduces biometric face scanning in ATMs. So, the research focuses on "implementing and enhancing security in ATM using facial recognition".

Table 6.1 Comparison of different ATM security models.

SL No.	Author (year)	Paper title	Technique used	Gap identified
1	Srivatsan Sridharan (2014)	Improvising authenticity and security of automated teller machine services.	This system provides the following facilities: withdrawing currency at any remote terminal, verifying the end user's identity using a PIN and an authentic one-time-Passkey (Pk) validation through a mobile.	The whole idea revolves around having a mobile, but in case the mobile is stolen, it will be a significant vulnerability for the user.
2	Priyanka Hemant Kale (2019)	Design of embedded based dual identification ATM card security system	In this paper, two options are included one- time password (OTP) and fingerprint detection for a successful transaction. The global system for mobile (GSM) is used for the generation of the OTP, and the OTP is sent to the mobile number to which the bank account is linked. The fingerprint of the user must be linked with the bank account so that the unique pattern of the fingerprint can be used to make a successful transaction.	A significant drawback of this paper was that they gave the users two ways to access the ATM instead of adding a security layer. The possibility of the intruder using the stolen mobile phone to access the user's account is still there.

(*Continued*)

Table 6.1 Continued

SL No.	Author (year)	Paper title	Technique used	Gap identified
3	Sayan Hazra (2019)	Smart ATM service	No ATM card is required for transactions in this suggested model; facial recognition, coupled with an aliveness check of the face, fingerprint verification, and OTP verification security checks have been considered. Cascade classifiers based on Haar features are used to detect faces.	In this proposed system, a biometric scan is only available for the account owner. Any other sub-user of the account must follow the OTP based transaction, which again is a considerable risk as the owner might not know who might be using the account if the mobile phone is stolen.
4	Gagan L. Choudhury (1996)	Squeezing the most out of ATM	The MAP/G/1 technique is utilized to provide the grounds for better calculations. Every important conclusion uses Poisson as a reference point. The effective bandwidth approximation is broken down using many independent sources.	Many assumptions and approximations were considered, which impacted the final output of the calculations. The Poisson approach, which is older, has been employed.

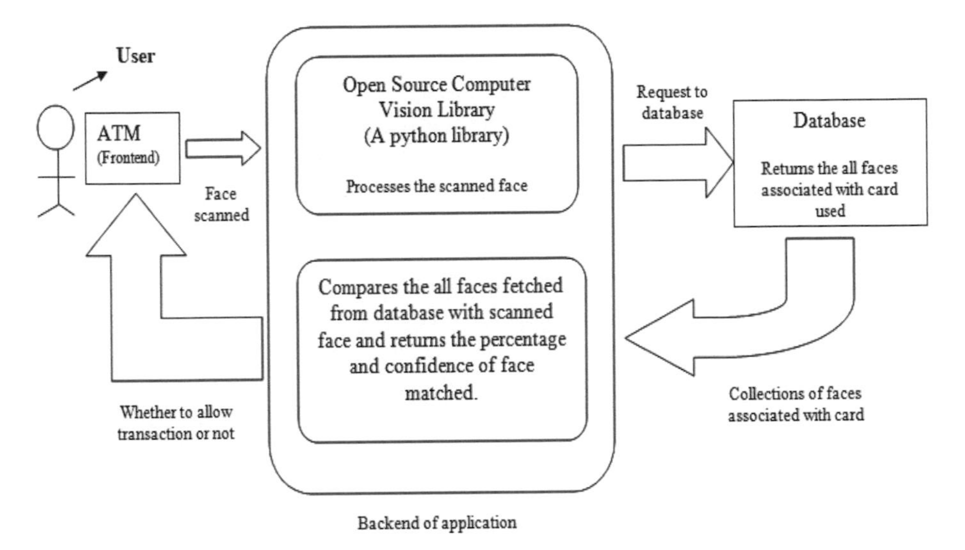

Figure 6.1 Structure of the proposed system.

6.4 Proposed System

When users insert their card in the slot, associated data is fetched from the database. If it fails to identify the information, the machine will notify the account holder that their PIN is probably compromised. In parallel, the user undergoes a face-scanning procedure. On scanning, the machine will match faces with family members of the cardholder. If the result of such matching is a "success", it will allow them to withdraw the money.

A Python library (OpenCV-python), which ships with pre-built binaries for various OpenCV machine learning algorithms, processes the user's captured face. In Figure 6.1, it is shown that the user interacts with the camera-equipped ATM. Meanwhile, on receiving a request for obtaining details, the database (MongoDB) sends all the faces associated with the used card back to the backend of the ATM. Once the faces are received, those faces undergo comparison with a captured face. Suppose any face comparison result is less than the matching parameter threshold (face distance), in that case, the user is authenticated and allowed to proceed with further interaction.

6.5 Methodology

1. **Insertion of card**: At first, the user (the person who uses an ATM card to withdraw cash from the ATM) will insert their card and allow the ATM to read the card. Behind the scenes, the card number is stored

and made ready to query the database. Meanwhile, the camera is made ready to capture.

2. **Scanning of face**: The user will stand in front of the camera, which will be installed on the ATM machine to capture the user's face. (In this model only, a webcam will be used.) The face image is temporarily stored on a disk to be processed and compared.

3. **Processing of face image**: Now, the system, with the help of the Open-Source Computer Vision Library, will process the scanned face of the user. Face detection using the "Haar cascades algorithm" is a machine learning-based approach where a cascade function is trained with input data. OpenCV already contains many pre-trained classifiers for face, eyes, smiles, etc. In this stage, the face is cut from the frame, resized and normalized before building the model.

4. **Getting the faces from a database**: In parallel, a secured and encrypted request is made through TCP to the database, which will return all faces associated with the card used. The database contains faces of all family members linked with a single card. The face is not stored as an image as doing so is highly inefficient and makes the database slower. Instead, the image of a face is stored on a disk, and the path address or location of these images is stored in a database which can be used to point to the stored face-on disk. After receiving the database's data (which contains a list of locations where the image is stored), the machine will try to fetch face images from these locations and then compare them with scanned faces with all available faces. Before comparing, these faces also undergo processing as in the previous step so that comparison is easy and produces more fruitful results.

   For example, let F_1 be the face scanned and let $(G_1, G_2, G_3 \ldots G_n)$ be the list of faces obtained from disk retrieved from the response of the database containing a list of path addresses or locations $(L_1, L_2, L_3 \ldots L_n)$. Then the comparison happens as shown below:

   $F_1 \times G_1 \rightarrow 0.85$
   $F_1 \times G_2 \rightarrow 0.94$
   $F_1 \times G_3 \rightarrow 0.21$

   .

   .

   .

   $F_1 \times G_n \rightarrow 0.54$

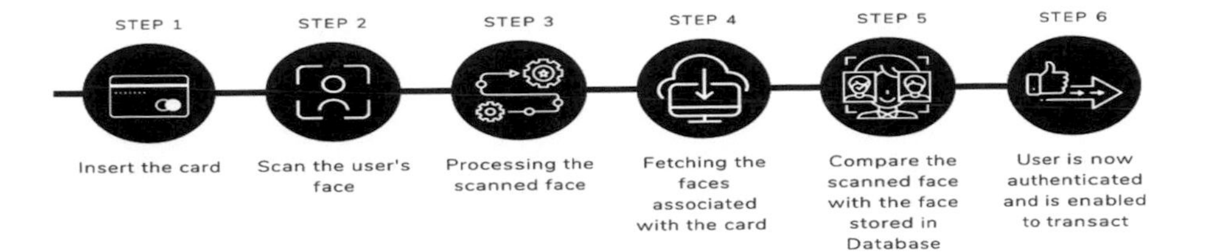

Figure 6.2 Operation of the proposed system.

where × represents comparison and the associated number is the comparison result.

5. **Comparison of face**: In the previous stage, comparison results in differences in the scanned face against each stored face in the database as a numerical value stored as an array. While trying to get the minimal difference out of all differences in array and mark that as best matched face, a user whose face was marked as best matched is trying to access the ATM (remember, there can be many faces, including family members) associated with a card, so best matched is the member of a family trying to access the ATM!). If the comparison meets the threshold, then for this best-matched face, the user is authenticated. If it fails (does not meet the threshold), the machine will ask for re-scanning the face and repeating the same procedure. Suppose it still fails for the third time in a row. In that case, the machine will notify the cardholder about the probably compromised card and prevent the transaction. The scanned face is also stored then in case of identifying the user if legal criminal case against them.

6. **Allowing operation**: Finally, after a user is authenticated, they are allowed to interact with their account or transact. Figure 6.2 depicts an overview of the proposed system.

6.6 Result and Discussion

As shown in Figure 6.3, aPIN-based transaction gives around 75% security as the validation of PIN is done through a mobile phone which, on being stolen, will become a significant vulnerability to the users, or if the user has lost their mobile phone, then the user will not be able to withdraw cash from the ATM. This technique was proposed by Srivatsan Sridharan (2014) in "Improvising Authenticity and Security of Automated Teller Machine Services ".

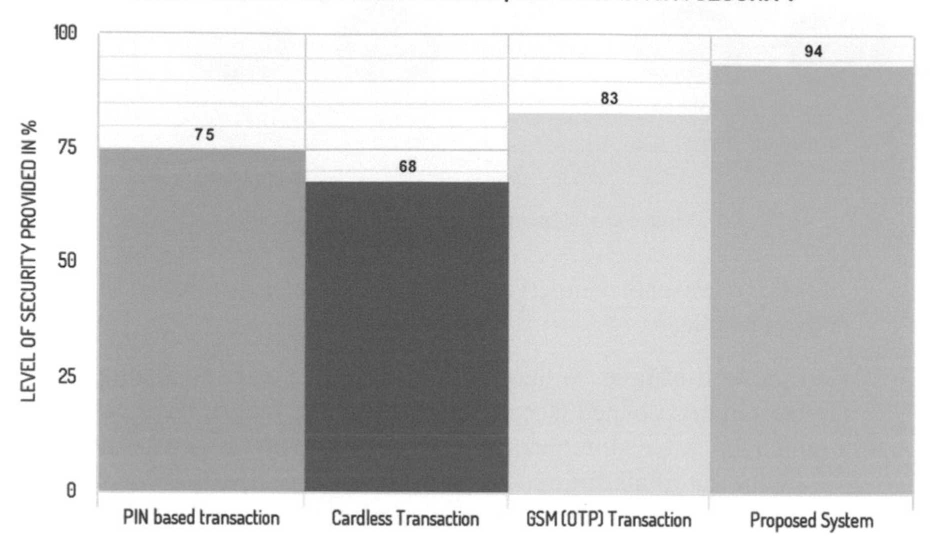

Figure 6.3 Comparison of different techniques used in ATM security.

Priyanka Hemant Kale (2019), in "Design of Embedded Based Dual Identification ATM Card Security System", proposed that for authentication, the user must either obtain an OTP or fingerprint detection for a successful transaction which gives the possibility of theft via stolen mobile phone, successfully avoiding the use of biometric scanning which could have caught intruder [18].

Later in 2019, Sayan Hazra proposed a "Smart ATM Service". He promoted the idea of card-less transactions by using biometric scans and OTP-based transactions, which were very innovative. However, it also had its flaw; the biometric scan was only available to the owner of the account and no one else. So, it made it hard to detect an intruder as the sub-users had to go through the old OTP based transaction, which has its risks [19-20].

Now, in the proposed system, we promote the idea of adding a layer of security to the existing system without removing any other layers of security. In this system, a database is created that stores all the faces of the owner and sub-owners of the account. It makes it compulsory for the users of the ATM to go through a biometric scan, as in [19-20]. Facial recognition, hence, immensely reduces the risk of money being stolen. Also, even if the intruder successfully steals the owner's mobile phone, they will not be able to withdraw any money from the ATM as their face does not match the faces stored in the database. The user will first insert their card and stand in front of the camera so that the user's face can be scanned and then processed using

OpenCV. Before processing, the system will ask the database to return all the faces associated with the card. The face will be processed and compared with the stored faces in the database to authenticate the user. Finally, the system will grant the user access to enter the pin.

Many scenarios can occur when a person attempts to use the card to access the ATM. The following section covers almost every scenario imaginable. It goes into great detail about how the function will be implemented on both the backend and frontend. The proposed system is about how the user's face is matched with image data stored in the database and whether or not the user is prompted for a PIN based on the results.

Case 1: Owner of the card accessing the ATM

The facial recognition procedure begins as soon as the card number is entered into the console, searching for the matching image data from a database. The system stores information such as his/her name, image information required to match the face, and family data in our database. The card number serves as the primary means of identification in this case. The cardholder is using the ATM, so after the faces are matched, the user is authenticated, and he or she is asked for the PIN. Figure 6.4 shows the owner login authentication page.

Case 2: Family member of card owner trying to access the ATM

Unlike the previous case, the original cardholder is not attempting to access the card, instead a family member is attempting to do so. Because their family

Figure 6.4 Owner login authentication.

```
_id: ObjectId("60b9ee2e2715c875381cf52b")
no: "234567891"
image: Array
  0: Object
    name: "Ayush"
    location: "image/AYUSH.jpg"
  1: Object
    name: "Sister"
    location: "image/SISTER.jpg"
```

Figure 6.5 ATM card's metadata.

Figure 6.6 Family members login authentication with same ATM card.

member's information is kept in the database, it is authenticated. He or she is asked for the PIN. So, after the card number is read, the picture recognition process begins, first looking for the direct cardholder image and then moving on to the next image stored in the database for recognition for the same card. Figure 6.5 depicts the ATM card metadata.

Family members also can use facial recognition technology to use the ATM services. Figure 6.6 depicts the family members login authentication with the same ATM card.

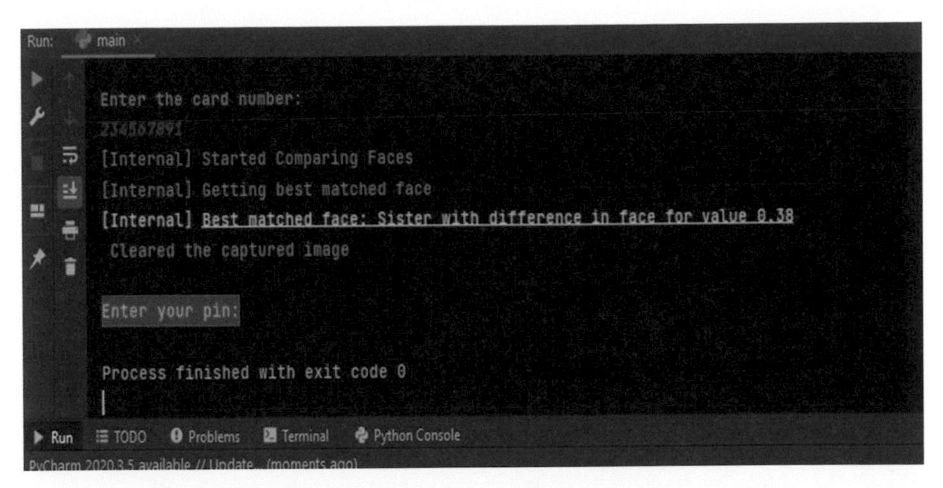

Figure 6.7 Family member of the card owner accessing the ATM.

After image verification is completed, the family members must enter the PIN to use ATM services. Figure 6.7 shows a family member of the card owner accessing the ATM.

Case 3: Intruder trying to access the ATM

Figure 6.8 depicts intruders trying to use the ATM services. In this case, an intruder is attempting to access the ATM. The intruder enters the correct card number, the request is sent to our database, and the face recognition procedure begins. However, their face is not matched with the cardholder's or family members' image data in this situation. As a result, the intruder is not prompted to enter the PIN, instead a warning is displayed, stating that the intruder's face does not match those in our database. Figure 6.9 shows the how the intruder is denied access to the ATM services.

6.7 Future Scope

AI technologies can help boost revenues through increased personalization of services to customers (and employees) and also lower costs through efficiencies generated by higher automation, reduced errors rates, and better resource utilization and uncover new and previously unrealized opportunities based on an improved ability to process and generate insights from vast troves of data.

The ability to predict if the cash rate being withdrawn equals what the cash management system predicted, i.e., will the ATM run out of cash? Many things can impact this. For example, if a nearby competing ATM fails, is that

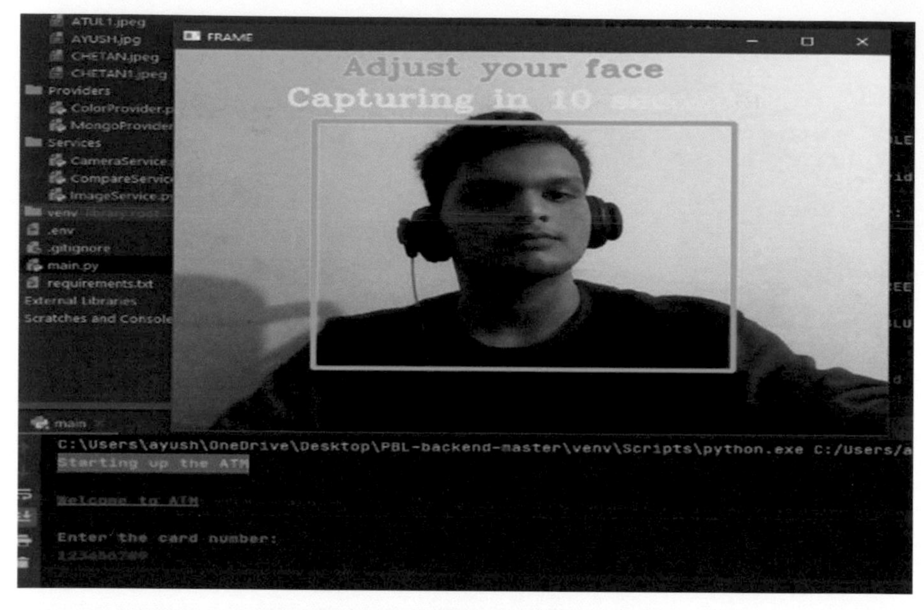

Figure 6.8 Intruder accessing the ATM.

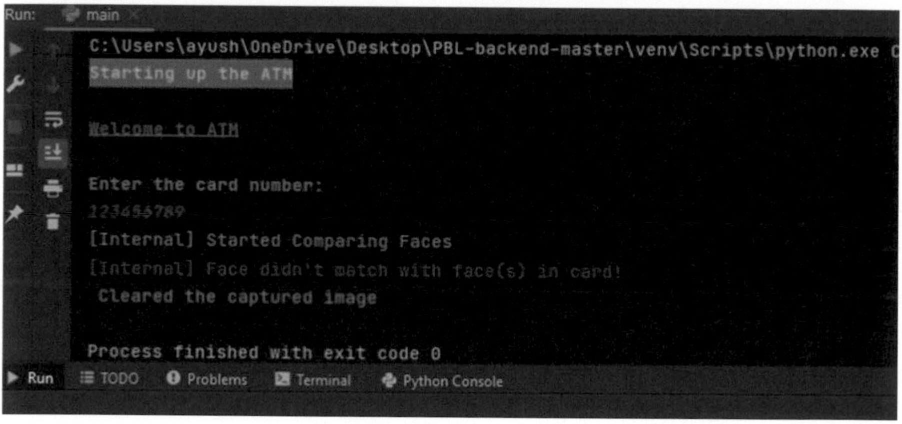

```
Run:    main
C:\Users\ayush\OneDrive\Desktop\PBL-backend-master\venv\Scripts\python.exe C
Starting up the ATM

Welcome to ATM

Enter the card number:
123456789
[Internal] Started Comparing Faces
[Internal] Face didn't match with face(s) in card!
 Cleared the captured image

Process finished with exit code 0
Run   TODO   Problems   Terminal   Python Console
```

Figure 6.9 Intruder failed authentication of ATM.

generating more traffic? Is the weather an influence on this ATM? Are the smaller denomination notes available on this ATM? If not, is this causing more significant withdrawal amounts? Together with predicting future values, these factors are what machine learning excels at.

Another use of AI is controlling the customer experience. Because AI algorithms will tend to store data sequences and history, this can be used for recommendation engines at the ATM. We could use this in the ATM experience to target advertising messages, ensuring the adverts which get the best results are targeted towards customers. Focus is also applied to reviewing how AI can be used to create custom experiences for every customer based on the behavior at ATMs.

The Cloud can aid in the centralization of services and the secure backup of sensitive data. The Cloud also meets the safety and security challenges of the field through highly secure protection and access and the possibility of redundancy of the data and applications it hosts.

As biometric authentication can include facial recognition or fingerprint technology, the interplay of the customer and ATM itself during a biometric authentication process requires localized processing. Whether an enhanced user experience, greater security, or increased functional elements, an edge computing implementation can contribute to these realizations.

6.8 Conclusion

In this chapter we highlighted the importance of ATM security and how there is a need for improvement in the transaction environment. There has been an upsurge in ATM fraud incidents such as card shimming, card skimming, and card trapping. As a result, it is necessary to devise a solution to address the current system's security concerns. An extra layer of security via facial recognition is done for authentication as it is more secure than traditional pin authentication. This will aid the ATM system in determining whether or not the correct person is using his or her card. Further, this will also help in digital forensics as intruder faces will be captured, reducing fraud. According to the statistics, HOG-based face recognition is very accurate requires less computation time and less storage space as trainee images are stored in the form of their projections on a reduced basis. We also concluded that facial recognition resulted in the highest accuracy against various other ways of the user trying to interact with ATMs. We also found that this model shows the qualitative analysis of algorithms used based on the metrics of existing algorithms.

The project software will not be limited for this specific use; it can be extended to use for cloud, blockchain, deep learning, etc., to make the sensitive environment more secure.

References

[1] Sayan Hazra (2019, March). Smart ATM Services. In 2019 Devices for Integrated Circuit (DevIC), 23–24 March, 2019, Kalyani, India.

[2] Egho-Promise Ehigiator Iyobor, Bamidele Ola, Emmanuel Opoku Sarkodie , A CYBER FORENSICS STUDY OF ATM DATA TRAFFIC, International Journal of Innovative Research in Information Security (IJIRIS) ISSN: 2349-7017 Issue 07, Volume 7 (August 2020) Gagan L. Choudhury (1996, February). Squeezing the Most Out of ATM. In IEEE Transactions on Communications, Vol. 44, No. 2, February 1996.

[3] Priyanka Hemant Kale, Dr. K. K. Jajulwar (2019). Design of Embedded Based Dual Identification ATM Card Security System. In 2019 9th International Conference on Emerging Trends in Engineering and Technology - Signal and Information Processing (ICETET-SIP-19).

[4] Srivatsan Sridharan, Gorthy Ravi Kiran, Sridhar Jammalamadaka (2014, February). Improvising Authenticity and Security of Automated Teller Machine Services. In IJCSMC, Vol. 3, Issue. 2, February 2014.

[5] Rupinder Saini, Narinder Rana, 'Comparison of various biometric methods', Institute of Engineering and IT, International Journal of Advances in Science and Technology (IJAST) Vol 2 Issue I (March 2014).

[6] Devinaga, R. (2010). ATM risk management and controls. European journal of economic, finance and administrative sciences. ISSN 1450-2275 issue 21.

[7] Anil K. Jain and Arun Ross. Introduction to Biometrics. In Anil K. Jain, Patrick Flynn, and Arun. A. Ross, editors, Handbook of Biometrics. Springer US, 2008.

[8] Ekenel HK, Stallkamp J, Gao H, Fischer M, Stiefelhagen R, Face Recognition for Smart Interactions, interact Research, Computer Science Department, University at Karlsruhe.

[9] Sayan Hazra (2019, March). Smart ATM Services. In 2019 Devices for Integrated Circuit (DevIC), 23–24 March, 2019, Kalyani, India.

[10] Pavan S. Rane, Prashant P. Sawat, Sourabh B. Shinde, Prof. Nitin A. Dawande (2018, June). ATM SECURITY. In International Journal of Advance Engineering and Research Development Volume 5, Issue 06, June -2018.

[11] N. Bansal and N. Singla, "Cash withdrawl from ATM machine using Mobile banking", Int. Conf. Computational The. Inform. And Communication Tech. (ICCTICT) India, pp. 535–539, March 2016.

[12] C. Courcoubetis, G. Kesidis, A. Ridder, J. Walrand and R. R. Weber, "Call acceptance and routing in ATM networks using inferences from measured buffer occupancy", IEEE Trans. Commun., vol. 43, pp. 1778–1784, 1995.

[13] Chavan Jagruti Kailas, Choudhary Kusum Savaram and Gavade Ankita Vijaykumar, "ATM security based on Iris Recognition", International

Research Journal of Engineering and Technology (IRJET), vol. 5, no. 4, Apr-2018.

[14] Christiawan, B. A. Sahar, A. F. Rahardian, and E. Muchtar, "Fingershield ATM – ATM Security System using Fingerprint Authentication," Int. Symposium Electronics and Smart Devices (ISESD) Indonesia, January 2019.

[15] Vinay Hiremath and Ashwini Mayakar, Face Recognition using Eigenface approach.

[16] P. Viola and M. Jones, "Rapid Object Detection using a Boosted Cascade of Simple Features", Proc. IEEE Comp. Soc. Conf., vol. 1, pp. 1-1, December 2001.

[17] G. Bradski and A. Kaehler, "Learning OpenCV: Computer vision with the OpenCV library" in O'Reilly Med. Inc., USA, 2008.

[18] R Aruna, V Sudha, G Shruthi, Rani R Usha and V Sushma, "ATM Security using Fingerprint Authentication and OTP", International Research Journal Of Engineering And Technology (IRJET), vol. 5, no. 5, May 2018.

[19] K. Baburao, M. K. Kishore, K. Nitya and K. Sunil Kumar, "Fingerprint and Iris Biometric Controlled Smart Banking Machine Embedded With GSM Technology For OTP", International Journal of Scientific Research and Review, vol. 7, no. 10, 2018.

[20] Sezin Kaymak, "Enhanced Principal Component Analysis Recognition Performance".

7

Network Forensics Architecture for Mitigating Attacks in Software-defined Networks

Immanuel Johnraja Jebadurai[1], Getzi Jeba Leelipushpam Paulraj[2*], Jebaveerasingh Jebadurai[3], and Salaja Silas[4]

[1,4]Professor, Department of CSE, Karunya Institute of Technology and Sciences, India
[2*]Associate Professor, Department of CSE, Karunya Institute of Technology and Sciences, India, getzi@karunya.edu
[3]Assistant Professor, Department of CSE, Karunya Institute of Technology and Sciences, India

Abstract

A software-defined network (SDN) decouples the control and data planes, enabling flexible functioning and management of networks. The programmable controller in an SDN has global information about the network, and it controls the network functionality. This provides high flexibility, scalability, and reliability to the network's operations. However, centralized control, the complex interaction of controller and devices, the communication process between control and data plane, and the lack of adequately devised security defence rules make it vulnerable to various types of attack. This chapter discusses the aspects of SDNs, attacks, and mitigation techniques towards securing SDNs. It also analyses the application of network forensics in SDNs. A network forensic-based framework has been devised to mitigate attacks in a software-defined network. Experimental analysis proves that the proposed framework could detect the attacks and provide a secure SDN.

7.1 Introduction

Traditional networking devices are configured using pre-defined commands based on networking operating systems. However, this limits the operation's flexibility and makes network management a challenging one. In modern infrastructures such as data centre networks, vehicular networks, and the Internet of Things, more flexibility is expected in load balancing, proactive fault tolerance, intelligent migration, and security[1]. Software-defined networks (SDNs) offer flexible networking by separating the data and control planes. The control plane is separated and mounted as a central controller gathering global information about the network, optimizing the network based on this information, and diffusing this information to individual network switches for forwarding the data packets. This reduces bandwidth overhead due to routing control messages exchanged between the network switches[2].

The network is revolutionized using an SDN. There have also been security threats against this architecture. An SDN mainly relies on software codes, making it more vulnerable to code injection-based attacks. SDN controllers exchange messages with the network switches communicating optimized routing strategies, which has enabled attackers to inject routing-based attacks. In addition to this denial of service attack, man in the middle attacks and masquerading attacks have also been attempted on the controller[3]. Hence it is important to detect possible attacks on the network and take suitable mitigation actions.

Network forensics is a technique that captures network data and analyses the data to identify attacks in the network. As there is a centralized controller in software-defined networks, the network forensic functionality can be incorporated into the SDN to investigate the network's security threats [4]. This chapter proposes a network forensic-based framework for detecting threats in the network. The remainder of this chapter is organized as follows: Section 2 explains the planes in software-defined networks. Section 3 discusses various attacks on SDNs. Section 4 describes the network forensic-based SDN controller for detecting and mitigating the attack. It also analyses the implemented forensic framework in a simulated environment. Section 5 concludes the chapter with future scope.

7.2 Software-defined Networking Planes

Network switches have three planes viz., data plane, control plane, and management plane[5].

- **Data plane:** The data plane has the input and output ports. Every packet reaches the data plane through the input port of the switch. The packet destination is found through the packet header examination and the referral to the forwarding table. After receiving the next-hop information from the table, the packet header is sealed again, and the packet is placed in the output port.
- **Control plane:** The responsibility of the control plane is to keep an updated forwarding table. Initially, the control plane runs the routing protocols to exchange messages between the network switches and compute the forwarding table. The control plane maintains up-to-date connectivity of the topology for error-free data transmission.
- **Management plane:** The network administrator uses a protocol such as the simple network management protocol (SNMP) to monitor the proper functioning of the network switches by collecting information through this plane.

Initially, in traditional network switches, packet forwarding is solely based on the destination address[6]. Then, it is improvised by incorporating policy-based routing where forwarding traffic from certain destinations is segregated in the form of flow, and they are treated uniformly. Then to improve the network's security, an access control list is maintained that suitably forwards the traffic based on the ports and the address information[7]. In most network cases, traffic is quarantined using specialized devices such as firewalls. The intrusion detection system is a forerunner for central software-based architecture[8].

An SDN aims to determine optimal routing design in the forwarding table by understanding the demand of the network and resilience to failures. Hence, the control plane is removed from network switches and moved to a central location to understand the complete view of the network at any time to take optimal decisions. The difference between the traditional and software-defined is depicted in Figure 7.1.

The SDN has a central controller that collects connectivity information from network switches and creates an optimal forwarding table for the switches. They are connected using southbound application programming interfaces (APIs). The data plane operating in the network switches receives the packet, refers it to the forwarding table and transfers the packet to the appropriate output port. The northbound API offers applications to the SDN through the application layers. The controller implements applications viz., load balancing, traffic analysis, and policies based on the configured

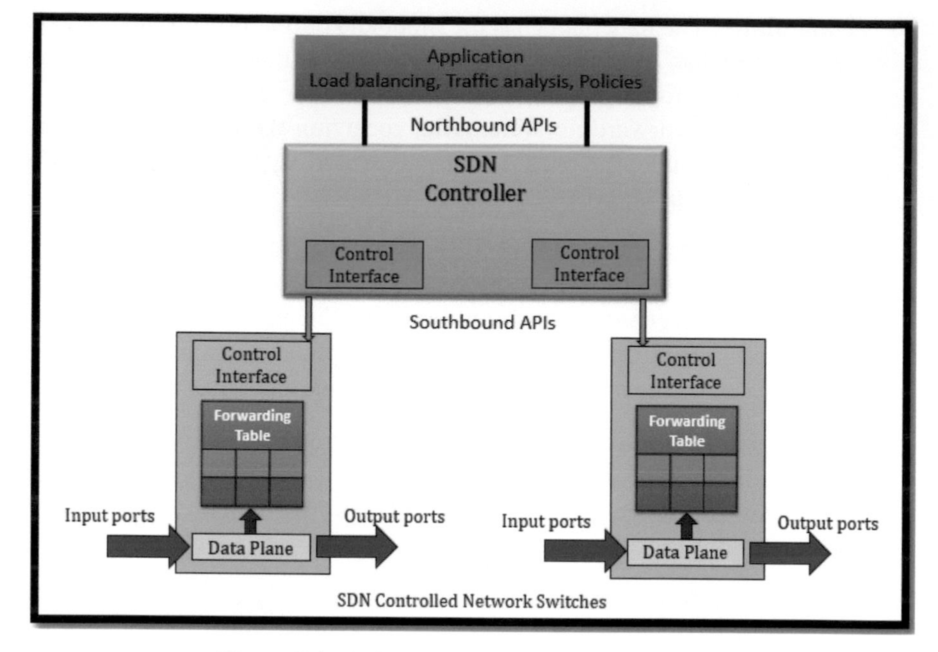

Figure 7.1 Software-defined network architecture.

application layer. By moving the control plane to the central controller, an SDN offers more flexibility, reduced cost, and improved profit. The SDN is also adopted and applied in various networks, data center networking, Internet of Things networking, and wireless networking[9]–[12].

7.3 Attacks in Software-defined Networks

As the SDN has promised various advantages compared with traditional networking, it is essential to analyze the security threats and implications in this centralized architecture. As the SDN controller is a controlling authority and the networking switches abide with the forwarding table formulated by the controller, it has become an point for attackers[3], [13]. This section discusses various threats in the SDN environment.

- An SDN environment is prone to IP and ARP spoofing attacks. The attacker can spoof MAC and IP address to receive updates from the SDN leading to attacks such as traffic analysis and denial of service attacks [3], [14].
- A distributed denial of service attack can be executed on the controller preventing it from communicating with the network switches. This

attack can be executed through UDP flooding, TCP flooding, and SYN flooding[15].

- A flow table overflow attack is hard to detect and affects the network switches' data plane. The SDN hardware switches maintain addressable memory to store the forwarding table. The attacker creates a record of flow entries and continuously sends them to network switches. This occupies the memory space and overflows the switch forwarding table. This brings down the performance of the network[16].
- A botnet with command and control mechanisms is a serious threat in a flow-based environment like an SDN. An infected entity compromises many devices to form a network of attackers, causing severe damage to the network[17].
- SDN controllers capture the overall picture of the network and create the forwarding table. The network switches communicate their link connectivity status to the controller. Having this information, the controller calculates the forwarding table. However, compromised nodes can send fake link information leading to a link discovery attack that results in incorrect forwarding table calculation[18].
- A crossfire attack aims at identifying pivotal links that connect the target zone to the rest of the network[19]. The attacker sends low-intensity traffic through pivotal links that eventually blocks the genuine traffic for the target zone. The target zone is disconnected as the switches in that zone is prohibited by the attacker in receiving the updated forwarding table.
- Various targeted attacks viz., implementation, enforcement, and policy attacks targeting northbound API, southbound API, data plane, and control plane have been discussed in [20].

The attacks discussed cause severe damage to the centralized control network and degrade the network's performance. These attacks have to be identified and disassociated from the network.

7.4 Network Forensics Architecture for Securing an SDN

In order to mitigate attacks on an SDN, a network forensic based architecture has been proposed. Network forensics is a science that deals with monitoring and analyzing real-time computer network traffic for possible cyber-attacks. The process begins with the identification stage in which abnormal network behavior is monitored through network indicators. If any anomaly is identified, data collection is initiated in that network location. Network activity in the identified location is monitored and logged. The data accumulated is suitably analyzed using algorithms to identify the type of attack and attacker. The final step is reporting and remedial measures. The attacker is suitably detached from the network for the proper functioning of the network[4], [21]. The proposed SDN network forensics has the following highlights.

- The capability of monitoring network and logging activity from various zones.
- Deep learning enables the analysis module to identify various attacks viz., denial of service, link discovery, crossfire, and table overflow attacks.
- A classification module for detaching the attacker from the network.

The proposed SDN architecture divides the network into three zones viz., network switches zone, controller zone, and application zone. SDN network forensics has various phases, viz., identification, data collection, data analysis and mitigation. Thc following section provides more insights into these phases.

7.4.1 Identification phase

During the first identification phase, the network behavior is monitored in two zones. This enables distributed monitoring of the network. The features collected during the identification phase are listed in Table 7.1.

Logs are maintained in every zone and communicated to the controller for further analysis. Log transfer follows a secure four-way handshake based method for enforcing confidentiality, integrity and authentication of the log data.

7.4.2 Data collection phase

The data collected at the devices and the nodes (M) must be transmitted to the controller (C) with confidentiality and authentication to ensure the integrity of the data. A four-way handshake methodology (FWHM) is proposed and implemented to provide confidentiality and authentication in the given environment. The proposed FWHM takes advantage of a public-key

Table 7.1 Data collected during the identification phase.

S. no	Zone	Features collected	Meaning
1	Controller zone	Source IP	Address of the sender
2		Destination IP	Address of the recipient
3		Flow ID	Data flow identifier
4		Flow length	Data flow size
5		Flow session duration	Data flow duration
6		Utilization rate	Bandwidth consumed
7		Resource usage	CPU and memory available
8	Network devises zone	Source IP	Address of the recipient
9		Flow ID	Data flow identifier
10		Flow length	Data flow size
11		Flow session duration	Data flow duration
12		Utilization rate	Bandwidth consumed
13		Resource usage	CPU and memory available
14		Flow duration	Data transmission time
15		Packet delivery ratio	Successful data transfer
16		Packet drop rate	Packet dropped
17		Response time	Time taken to forward a packet

cryptosystem. It has two phases: (i) exchange of public keys among them and (ii) exchange of data with confidentiality and authentication.

(i) Exchange of public keys

Initially, both M and C exchange their public keys one-to-one. That is, each node or device independently exchanges its public key with the controller. It is assumed that the controller is never compromised. A key authority service (KAS) is running in the C to exchange public keys. The following are the steps involved in this key exchange protocol.

Step 1: C shares its public key (PU_C) with KAS as plain text.

Step 2: M shares its public key (PU_M) with KAS as plain text.

This initial exchange does not require the keys to be encrypted or encoded as the further steps ensure the integrity of this message exchange.

Step 3: The KAS shares a ticket (T_C) encrypted by its private key (PR_{KAS}) containing the timestamp of third message (t_3), identification of C (IP_C) and the public key of C (PU_C) with the C where

- The timestamp (t_3) validates whether the ticket (T_C) is current.
- Identification of C (IP_C) confirms the IP address of the controller.

Step 4: The KAS shares a ticket (T_M) encrypted by its private key (PR_{KAS}) containing the timestamp of the fourth message (t_4), identification of C (IP_M) and the public key of C (PU_M) with the node M where

- The timestamp (t_4) validates whether the ticket (T_M) is current.
- Identification of M (IP_M) confirms the IP address of the node.

This exchange of the tickets holding the corresponding public keys ensures that the public keys can be exchanged among the nodes and the controller with confidence. Because the tickets T_C and T_M can be decrypted only by using the public key of the KAS (PU_{KAS}). It also confirms that the tickets were created only by the KAS, a trusted entity.

For controller (C), the ticket provided by KAS is of the form

$$T_C = E(PR_{KAS}, [\ IP_C \,||\, t3 \,||\, PU_C]).$$

Controller (C) shares this ticket (T_C) to the device (M). M utilizes the public key of KAS (PU_{KAS}) to decrypt the ticket T.

Step 5: C shares the ticket (T_C) with M.

Step 6: M shares the ticket (T_M) with C.

After steps 5 and 6, both the controller and the devices have successfully exchanged their public keys. The entire operation is given in Figure 7.2.

(ii) Exchange of data with confidentiality and authentication.
Once the device's public keys (M) and the controller (C) have been exchanged successfully, the data can be exchanged among them with confidentiality and authentication to thwart active and passive attacks on the communication setup. The data exchange happens between C and M through the following steps.

Step 7: The device (M) encrypts the following components using the public key of C (PU_C) and sends it to C. The components are M's identifier (IP_M) and a unique identifier (Q_M) of the device.

Step 8: The controller (D) encrypts the following components using the public key of M (PU_M) and sends it to device M. The components are the unique identifier (Q_M) of the device and the unique identifier (Q_C) of the controller.

The unique identifiers are generated by the controller and the device independently using the *system time* and a *pseudo-random number*. Hence, the adversary cannot deduce these unique identifiers.

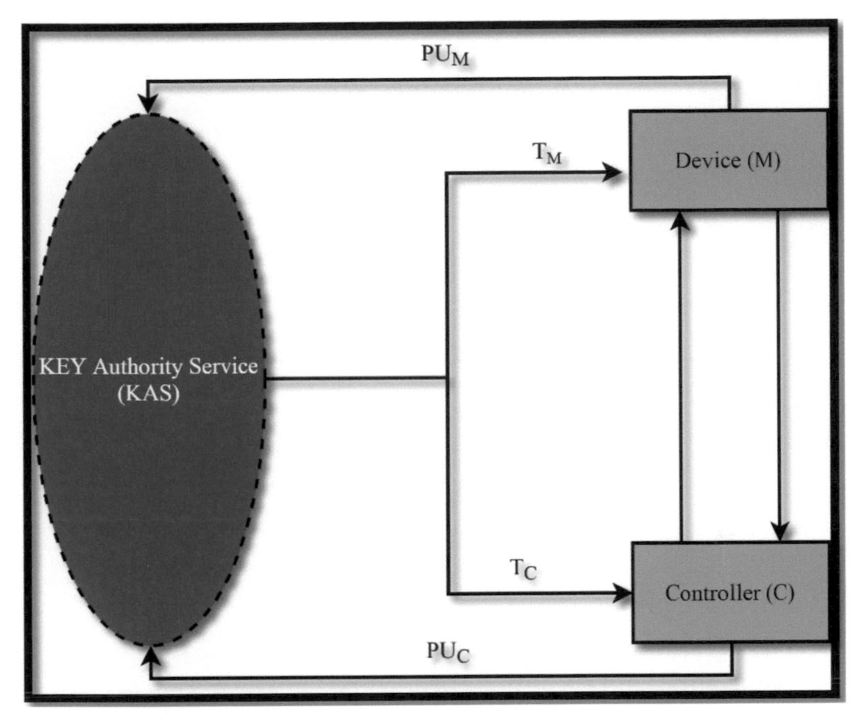

Figure 7.2 Exchange of public keys.

Step 9: The device (M) sends back the unique identifier (Q_C) of the controller encrypted by the public key of the controller (PU_C).

Step 10: The device (M) is now ready to send the collected data (X) to the controller (C) using the exchanged keys. The device (M) encrypts the data (X) using its private key (PR_M) and then by using the public key of controller C (PU_C). It is denoted by

$$E(PU_C, E(PR_M, X))$$

Because the data (X) is encrypted using PR_M, PU_M can be decrypted only. This proves that the data (X) has been sent by the device (M) only. Further, the data is encrypted by PU_C, restricting only the controller to decrypt and get the data (X).

The entire data exchange procedure is depicted in Figure 7.3. Once the initial public keys exchange is completed, steps 7–10 are repeated for different data exchanges.

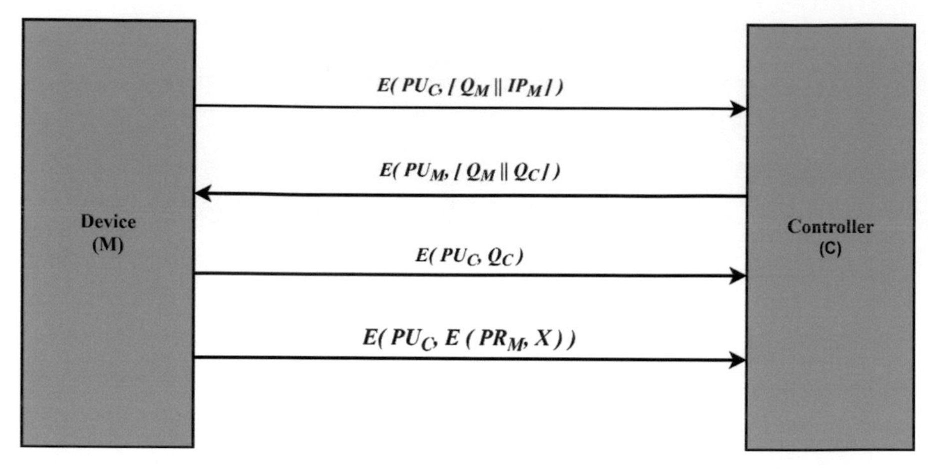

Figure 7.3 Data exchange with confidentiality and authentication.

7.4.3 Analysis phase

Analysis is the critical part of the network forensic architecture. The analysis can be handled by a third-party service or internally by the controller.

7.4.3.1 Detection of flooding attack

Since deep learning-based approaches have proved to show higher accuracy for multi-class classification[17], [22], the detection module in the controller implements these algorithms to identify the possible threats. The detection module detects the type of flooding attack using echo state networks. An echo state network (ESN) is an advanced, recurrent neural network overcoming the issues in a recurrent neural network. The ESN consists of three layers viz., input layer, a fixed reservoir, and an output layer. The input data is captured in real-time on the SDN testbed.

Let $S = \{S_1, S_2 .... S_n\}$ be the input data samples fed into the network as input. Let W_{in} be the weight between the input and the reservoir, W be the weight between the reservoir and the output layer and W_{out} be the weights between the reservoir and the output layer. The ESN architecture is shown in Figure 7.4.

The output of the reservoir hidden layer is shown in equation (7.1). The input has m' internal units, the reservoir holds K' units, and the output holds C internal units as the network has to classify C classes.

$$h_t = f\ (W_{in} S_t + W_{in} h_{t-1}) \tag{7.1}$$

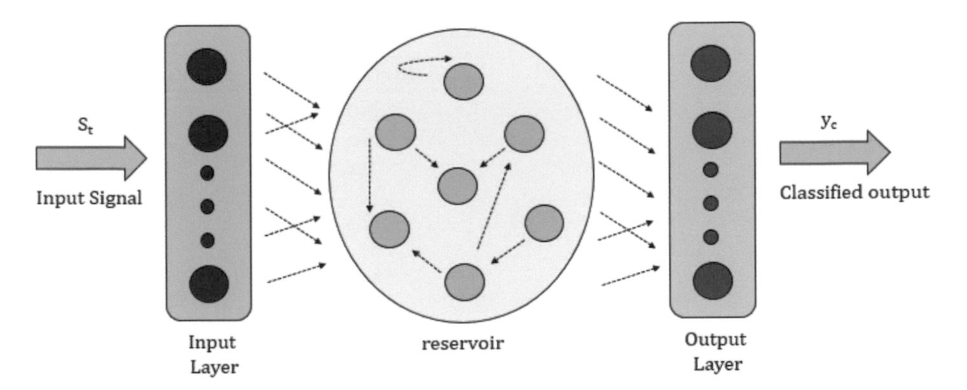

Figure 7.4 Echo state network architecture.

where h_t represents the hidden layer's output and $f(.)$ denotes the activation function. The output is denoted by equation (7.2).

$$y_t = f\left(W_{out} h_t\right) \tag{7.2}$$

where y_t denotes the classified output of the ESN.

7.4.3.2 Detection of a flow table overflow attack

A flow table overflow attack occurs in the switches. The attacker is assumed to send flow table entries randomly to selected switches in the network. Certain messages are collected from the switches. A number of flow entries newly added in every switch are captured and sent to the controller for analysis. The data is sent through a secure four-way handshake protocol and the flow statistics.

Let the entry addition rate be $r_i(t)$, where i denotes the ith switch in the network and t denotes the entry time t. As the controller holds the complete picture of the network, conditions such as a newly added switch, broken link, and topology changes are also monitored as given in Table 7.2.

Every condition has a permitted flow entry $p_i^c(t)$, denoting the permitted flow entry at switch i at condition c. The condition is checked by equation (7.3).

$$\begin{aligned} &\left(\text{if } (r)_i(t) > p_i^c(t)\right), && \text{Attack detected} \\ &\left((r)_i(t) < p_i^c(t)\right), && \text{No attack detected} \\ &\left(\text{if } (r)_i(t) = p_i^c(t)\right), && \text{Further investigation is required} \end{aligned} \tag{7.3}$$

Table 7.2 Checking conditions.

Condition	Broken link	Added switch	Topology change
1	No	No	No
2	No	No	Yes
3	No	Yes	No
4	No	Yes	Yes
5	Yes	No	No
6	Yes	No	Yes
7	Yes	Yes	No
8	Yes	Yes	Yes

When a switch is suspected to be a compromised node, suitable action is taken to detach the switch from the network.

7.5 Experimental Analysis

An experimental analysis has been carried out on a Mininet network emulator. Mininet version 2.3.0 was installed on Virtual Box 6.1.28 windows package[23], [24]. The experiment was done on an Intel i7 core processor with 16GB RAM. The topology used for experimentation is shown in Figure 7.5. Open v Switches controlled by a Pox controller is connected[25].

The topology has eight switches connected in two layers. The switches from s1 to s5 creates the access layer connecting the host to the network. Switches s6 to s8 connect the access switches to the controller forming the distribution layer as shown in Figure 7.5.

7.5.1 Performance analysis on flooding attack detection

The links operate with a bandwidth of 100 Mbps and 5 ms delay. Initially, emulation is performed for one hour by generating normal and abnormal traffic using TCP, UDP, and ARP[14], [26]. The indicators mentioned in Table 7.1 were collected from the host, switches, and the controller. 64,000 samples are finalized after the data cleaning. The collected data is analyzed using echo state networks. It is then compared with the convolutional neural network and recurrent neural network[27]. The training phase is carried out using 64,000 samples. Testing is performed by randomly initiating normal and abnormal traffic in the network. Performance metrics, accuracy, precision, recall, and f1-score have been measured[28], [29]. The output classes are regular traffic, TCP flooding, UDP flooding, and ARP flooding. Such flooding attacks may lead to denial of service for the user's application. The comparative analysis is shown in Table 7.3.

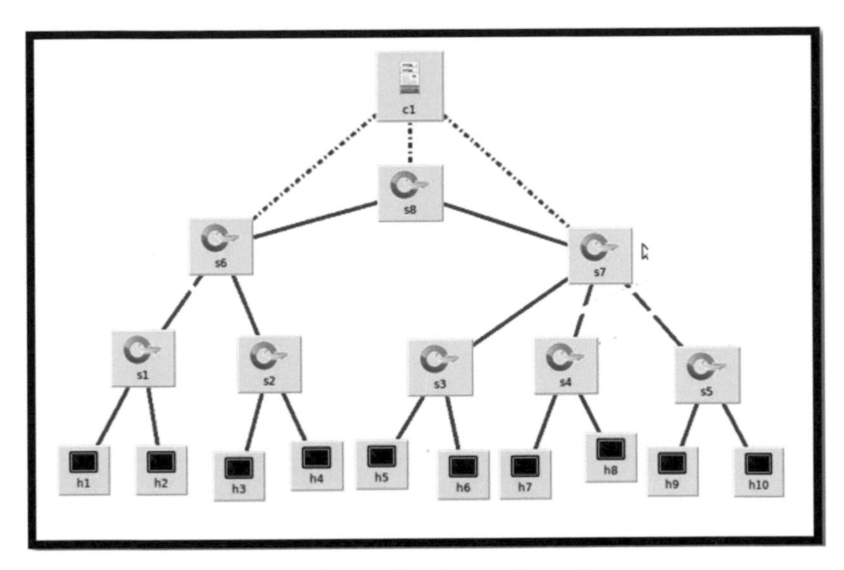

Figure 7.5 Emulated network topology.

Table 7.3 Comparative analysis of CNN, RNN, and ESN.

Type of attack		CNN	RNN	ESN
TCP flooding	**Precision**	0.84	0.86	0.91
UDP flooding		0.83	0.82	0.92
ARP flooding		0.85	0.90	0.91
Normal		0.86	0.87	0.92
TCP flooding	**Recall**	0.83	0.85	0.90
UDP flooding		0.82	0.83	0.92
ARP flooding		0.83	0.90	0.93
Normal		0.86	0.87	0.91
TCP flooding	**F1-score**	0.83	0.85	0.92
UDP flooding		0.84	0.81	0.93
ARP flooding		0.85	0.89	0.91
Normal		0.82	0.87	0.92
	Accuracy	0.88	0.86	0.92

The experimental results show that an ESN can identify a flooding attack with 92% accuracy. Even though an RNN shows better results in identifying ARP flooding attacks, it fails to detect TCP, UDP, and regular traffic. However, the ESN classifies all the traffic with better accuracy. By classifying flooding attacks, severe damage is caused due to denial of service attacks [30].

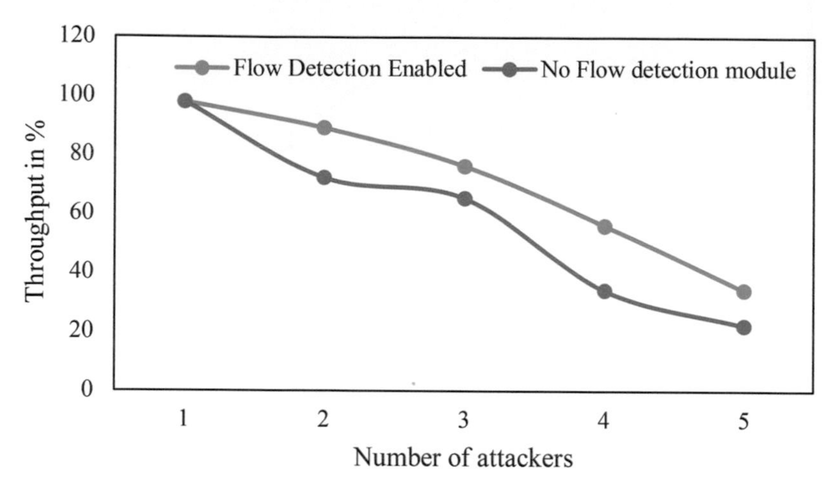

Figure 7.6 Throughput vs number of attackers.

7.5.2 Performance analysis on flow table overflow attack detection

Performance analysis is carried out by setting up an attack on the switches. The attack starts with a single switch scaling up to four switches. The packet delivery ratio has been measured. The performance analysis on packet delivery ratio by varying the number of compromised nodes is shown in Figure 7.6. Results are compared by running the flow table overflow detection module and not running the overflow module.

From Figure 7.6 it is observed that the throughput is relatively high when there are few attackers. Throughput decreases with an increase in the number of attackers. However, the throughput is relatively improved with the flow table overflow detection module. As the flow table detection module detects the compromised switch, the abnormal traffic stops for that switch, which keeps the throughput unaffected.

7.6 Conclusion

A software-defined network offers higher flexibility in the operation of modern computer networks. Its customizability, ease of innovation, and flexibility enable current technologies such as cloud data centers, wireless networks and the Internet of Things to adopt this for their networking infrastructure. Having the responsibility of handling massive data and applications in these networks, an SDN is prone to various attacks, viz., flooding, denial of service, and flow table overflow attacks. This has caused packet loss and delay in data delivery. Network forensics offers a secure solution by identifying, collecting, and analyzing network traffic. This work proposes a network forensic framework equipped with a deep neural network model for offering security. Data is collected in the host, switches, and controller. The collected data is securely transmitted to the controller using a four-way handshaking protocol. The data is used to train the echo state network model. The trained model identifies the possible flooding attack in the network, thereby hindering the denial of service attack. Performance analysis with CNN and RNN architectures shows that the ESN architecture has 92% accuracy. In future, the deep neural network parameters can be tuned for improved accuracy. More attack scenarios will be incorporated for a generalized secure forensic architecture.

7.7 Acknowledgement

This work was carried out in the Centre of Excellence for Networking Academy at Karunya Institute of Technology and Sciences, Coimbatore.

References

[1] M. Jammal, T. Singh, A. Shami, R. Asal, and Y. Li, "Software-defined networking: State of the art and research challenges," *Comput. Networks*, vol. 72, pp. 74–98, 2014, doi: 10.1016/j.comnet.2014.07.004.

[2] H. Farhady, H. Lee, and A. Nakao, "Software-Defined Networking: A survey," *Comput. Networks*, vol. 81, pp. 79–95, 2015, doi: 10.1016/j.comnet.2015.02.014.

[3] I. Alsmadi and D. Xu, "ScienceDirect Security of Software Defined Networks: A survey," *Comput. Secur.*, vol. 53, pp. 79–108, 2015, doi: 10.1016/j.cose.2015.05.006.

[4] R. Y. Patil and S. R. Devane, "Network Forensic Investigation Protocol to Identify True Origin of Cyber Crime," *J. King Saud Univ. - Comput. Inf. Sci.*, no. xxxx, 2019, doi: 10.1016/j.jksuci.2019.11.016.

[5] M. Chahal, S. Harit, K. K. Mishra, A. Kumar, and Z. Zheng, "A Survey on software-de fi ned networking in vehicular ad hoc networks: Challenges, applications and use cases," vol. 35, no. July, pp. 830–840, 2017, doi: 10.1016/j.scs.2017.07.007.

[6] R. Masoudi and A. Ghaffari, "Software de fi ned networks: A survey," Journal of Network and Computer Applications, vol. 67, pp. 1–25, 2016, doi: 10.1016/j.jnca.2016.03.016.

[7] S. Saraswat, V. Agarwal, H. P. Gupta, R. Mishra, and A. Gupta, "Challenges and solutions in Software Defined Networking: A survey," *J. Netw. Comput. Appl.*, vol. 141, no. May, pp. 23–58, 2019, doi: 10.1016/j.jnca.2019.04.020.

[8] C. Birkinshaw, E. Rouka, and V. G. Vassilakis, "Implementing an intrusion detection and prevention system using software-defined networking: Defending against port-scanning and denial-of-service attacks," *J. Netw. Comput. Appl.*, vol. 136, pp. 71–85, 2019, doi: 10.1016/j.jnca.2019.03.005.

[9] H. Mostafaei and M. Menth, "Software-de fi ned wireless sensor networks: A survey," Journal of Network and Computer Applications, vol. 119, no. June, pp. 42–56, 2018, doi: 10.1016/j.jnca.2018.06.016.

[10] Y. Jararweh, M. Al-ayyoub, A. Darabseh, and E. Benkhelifa, "Software defined cloud: Survey, system and evaluation," *Futur. Gener. Comput. Syst.*, vol. 58, pp. 56–74, 2016, doi: 10.1016/j.future.2015.10.015.

[11] K. Mathews, B. Betty, R. Malekian, and A. M. Abu-mahfouz, "Software defined wireless sensor networks application opportunities for efficient network management: A survey R," *Comput. Electr. Eng.*, vol. 66, pp. 274–287, 2018, doi: 10.1016/j.compeleceng.2017.02.026.

[12] K. Nisar *et al.*, "Internet of Things Review article A survey on the architecture, application, and security of software defined networking: Challenges and open issues," *Internet of Things*, vol. 12, p. 100289, 2020, doi: 10.1016/j.iot.2020.100289.

[13] J. Xie, D. Guo, Z. Hu, T. Qu, and P. Lv, "Control plane of software defined networks: A survey," *Comput. Commun.*, vol. 67, pp. 1–10, 2015, doi: 10.1016/j.comcom.2015.06.004.

[14] G.Darwesh,A.A.Vorobeva,V.M.Korzhuk,"Anefficientmechanismtodetect and mitigate an ARP spoofing attack in software-defined networks," vol. 21, no. 3, pp. 401–409, 2021, doi: 10.17586/2226-1494-2021-21-3-401-409.

[15] M. Myint Oo, S. Kamolphiwong, T. Kamolphiwong, and S. Vasupongayya, "Advanced Support Vector Machine-(ASVM-) based detection for Distributed Denial of Service (DDoS) attack on Software Defined Networking (SDN)," *J. Comput. Networks Commun.*, vol. 2019, 2019, doi: 10.1155/2019/8012568.

[16] S. Xie, C. Xing, G. Zhang, and J. Zhao, "A Table Overflow LDoS Attack Defending Mechanism in Software-Defined Networks," *Secur. Commun. Networks*, vol. 2021, 2021, doi: 10.1155/2021/6667922.

[17] F. Tariq and S. Baig, "Machine Learning Based Botnet Detection in Software Defined Networks," *Int. J. Secur. Its Appl.*, vol. 11, no. 11, pp. 1–12, 2017, doi: 10.14257/ijsia.2017.11.11.01.

[18] S. Sen Baidya and R. Hewett, "Link discovery attacks in software-defined networks: Topology poisoning and impact analysis," *J. Commun.*, vol. 15, no. 8, pp. 596–606, 2020, doi: 10.12720/jcm.15.8.596-606.

[19] A. R. Narayanadoss, T. Truong-Huu, P. M. Mohan, and M. Gurusamy, "Crossfire attack detection using deep learning in software defined its networks," *IEEE Veh. Technol. Conf.*, vol. 2019-April, 2019, doi: 10.1109/VTCSpring.2019.8746594.

[20] A. Shaghaghi, M. A. Kaafar, R. Buyya, and S. Jha, "Software-Defined Network (SDN) data plane security: Issues, solutions, and future directions," *Handb. Comput. Networks Cyber Secur. Princ. Paradig.*, pp. 341–387, 2019, doi: 10.1007/978-3-030-22277-2_14.

[21] L. F. Sikos, "Forensic Science International: Digital Investigation Packet analysis for network forensics: A comprehensive survey," *Forensic Sci. Int. Digit. Investig.*, vol. 32, p. 200892, 2020, doi: 10.1016/j.fsidi.2019.200892.

[22] R. Pradeepa and M. Pushpalatha, "Artificial neural network (ANN) based DDoS attack detection model on software defined networking (SDN)," *Int. J. Recent Technol. Eng.*, vol. 8, no. 2, pp. 4887–4894, 2019, doi: 10.35940/ijrte.B3670.078219.

[23] A. Mayssara A. Abo Hassanin Supervised, *International Conference on Communication, Computing & Systems*. 2014.

[24] R. L. S. De Oliveira, C. M. Schweitzer, A. A. Shinoda, and L. R. Prete, "Using Mininet for emulation and prototyping Software-Defined Networks," *2014 IEEE Colomb. Conf. Commun. Comput. COLCOM 2014 - Conf. Proc.*, no. June, 2014, doi: 10.1109/ColComCon.2014.6860404.

[25] R. ur Rasool, K. Ahmed, Z. Anwar, H. Wang, U. Ashraf, and W. Rafique, "CyberPulse++: A machine learning-based security framework for detecting link flooding attacks in software defined networks," *Int. J. Intell. Syst.*, no. April, 2021, doi: 10.1002/int.22442.

[26] Y. C. Lai, A. Ali, M. S. Hossain, and Y. D. Lin, "Performance modeling and analysis of TCP and UDP flows over software defined networks," *J. Netw. Comput. Appl.*, vol. 130, no. June 2018, pp. 76–88, 2019, doi: 10.1016/j.jnca.2019.01.010.

[27] S. Haider *et al.*, "A Deep CNN Ensemble Framework for Efficient DDoS Attack Detection in Software Defined Networks," *IEEE Access*, vol. 8, pp. 53972–53983, 2020, doi: 10.1109/ACCESS.2020.2976908.

[28] M. M. Raikar, S. M. Meena, M. M. Mulla, N. S. Shetti, and M. Karanandi, "Data Traffic Classification in Software Defined Networks (SDN) using supervised-learning," *Procedia Comput. Sci.*, vol. 171, no. 2019, pp. 2750–2759, 2020, doi: 10.1016/j.procs.2020.04.299.

[29] S. Sen, K. D. Gupta, and M. Manjurul Ahsan, "Leveraging Machine Learning Approach to Setup Software-Defined Network(SDN) Controller Rules During DDoS Attack," no. February, pp. 49–60, 2020, doi: 10.1007/978-981-13-7564-4_5.

[30] H. Wang, U. Ashraf, K. Ahmed, and Z. Anwar, "Journal of Network and Computer Applications A survey of link flooding attacks in software defined network ecosystems," *J. Netw. Comput. Appl.*, vol. 172, no. March, p. 102803, 2020, doi: 10.1016/j.jnca.2020.102803.

8

The Self-destructive Behavioural Effects of Virtual Addiction on Cyber Crime Scene Investigation of Victimless Crimes

V. Sabapathi[1], and J. Selvin Paul Peter[2]

[1]Research Scholar, Department of Computer Science and Engineering, SRM Institute of science and technology, Kattankulathur, Chennai-603 203, India
[2]Associate Professor, Department of Computer Science and Engineering, SRM Institute of science and technology, Kattankulathur, Chennai-603 203, India
Email: selvinpj@srmist.edu.in

Abstract

Virtual addiction-based activity analysis is challenging because of the vast amount of data exchange and the expansion of widespread smartphone use. The reality is that smartphones are a dynamic activity platform with multiple benefits that everybody utilizes, and people have become heavily reliant on them. A massive amount of personal data is kept on cell phones, which can be used to forecast user behaviour using log functions, and other keyword storing and mapping methodological approaches. Hackers can decode information and cause damage by examining the behaviour pattern. However, this is a third-party attack; our debate is about self-destructive conduct, such as virtual addiction, that impacts cybercrime. This study's primary goal is to examine how numerous addictions affect human cognition and cause painful personal health via virtual media-based addictive uses such as smartphones, virtual reality games, and television. This virtual consumption is strongly linked to emotional connection, human habit formation, and compulsive addictive behaviour. The study's primary insight is that virtual addiction is self-destructive, contributing to numerous victimless crimes. The original study premise is that when overwhelming addictive behaviours were investigated, it was found that illegal activity may result in self-destructive behaviour, such as attempted suicide, loss of money and possessions, biological and behavioural unfavourable cognition, anxiety, nervousness, and a lack of self-esteem.

8.1 Introduction

This study was inspired by the rapid rise in digital platforms, which has resulted in addiction-related self-destructive conduct and victimless crimes. Additionally, it is critical to developing cognitive paradigms to raise awareness regarding addictive behaviour concerns, since artificial intelligence in terms of cognitive model development has become influential in various intelligent healthcare systems. Numerous mental disorders have affected human cognitive processes in recent decades. After the evolution of smartphone technology, the world has become virtually connected rather than in direct communication due to the multiple features of smartphones, which have become more necessary. However, people have become stuck using smartphones, becoming virtually dependent on them. As per a previous study, virtual related multiple mental illnesses have been defined such as nomophobia. Nomophobia is a psychological illness and is defined as feeling uncomfortable without a mobile phone.

Moreover, multiple adverse cognitive illnesses have been identified, such as schizophrenia, and unipolar and bipolar depression. At the same time, smartphone use continues to become an addiction. These types of cognitive illness lead to poor academic performance, lack of focus, and low self-esteem. Due to the widespread use of smartphones, several cognitive illnesses, such as Alzheimer's disease, have arisen. Nomophobia and smartphone addiction are linked to comorbidities, according to much research. Eating disorders, obsessive-compulsive disorder, depression, and other behavioural addiction diseases (gambling, compulsive shopping, and so on) are all examples of behavioural addiction disorders [1–2]. The concern of (1) not being able to converse, (2) losing connectedness, (3) not being able to obtain information, and (4) giving up convenience were all validated [3]. The experience of not communicating echoes the loss of instant communication and the inability to use gadgets for instant communication. It is appealing to be physically alone but socially active on the internet. Due to the variety of platforms, people have become more dependent on virtual systems. Virtual systems are not limited to smartphones but extend to multiple gadgets and electronic devices such as virtual reality gaming and televisions. However, these types of inputs are cognitively prompting repeated use. This type of addiction-related disorder leads to multiple cybercrimes.

8.2 Related Study

It is still disputed whether or not a person is dependent on specific substances. Casting judgment concerning compulsive behaviour needs consciousness, considering the confusing nature of things. There is no widely accepted contextual measurement, as according to previous studies. According to observational studies, people with social phobias become dependent on communication through their mobile devices or computer to avoid direct social contact [4-5]. Social anxiety has also been discovered to be highly comorbid with depression, and it is frequently present in depressed people who do not fit the diagnostic criteria [6]. It is estimated that more than 90% of children and adolescents in the United States play video games, with a considerable number of individuals spending a growing amount of time doing so [7-8]. The gaming industry generated $155 billion in revenue by 2020. Analysts estimate the business will earn more than $260 billion in revenue by 2025. [9]. It should not be assumed that gaming is damaging in and of itself. On the other hand, video games may meet some psychological requirements in users, such as identity expression, a sense of mastery and achievement, and a need to escape from reality.

Despite some positive connotations, excessive gaming can lead to a variety of negative consequences (e.g., financial losses because online games frequently require money to continue playing, psychological detachment, sleep deprivation, eating and nutritional problems, a lack of personal and social interaction, depression, and anxiety[18]. Depression, anxiety, impulsivity, poor self-regulation, academic challenges, limited social engagement, and a lack of familial interaction are all side effects of excessive smartphone use[18-20]. Increased smartphone use has also been linked to lower physical activity and more sedentary behaviour, sleep disruptions, and various physical issues (neck stiffness, hazy vision, etc.)[21-23]. In the face of uncertainties, inversion evaluation questionnaires are not only the assessment of ambiguity but also the validation of the designer's application. In this context, inverse analysis was conducted on the subject of fuzzy arithmetic, culminating in flexible inverted mathematics that relies on technology. Furthermore, as shown in Table 8.1, numerous studies in cognitive healthcare science, investigative examinations, and other areas are always essential.

Table 8.1 Healthcare investigation gap.

Problem discussed in recent studies	Merits	Limitations
Analysis based on substance-related addiction.	Discussed more insights about the concern substance addiction.	Addiction to substances and consumables is studied, but cognitive dependency and behaviour tendencies are only studied to a limited extent [10].
Discussion of online gambling and gaming dependent pattern related addiction.	Several studies discussed the seriousness of online gaming gambling addiction patterns.	The cognitive association of online platform virtual arrest behaviour still requires more insight [11, 12].
Addiction behaviour patterns and strategies [13].	Substance and virtual dependents and cognitive illness are discussed separately.	It can incorporate and evaluate the cognitive associations of addiction in both (substance, virtual platforms) contexts.
Addiction related preventive measures discussed with medical technology and ML, AI assistance.	The technology advances dramatically assist in the post-treatment process.	The natural intelligence interpretations and applications in pretreatments require numerous associations among the cognitive relationships [14, 15].
Analysis of an uncertain context of human behaviours has significant challenges [16].	There are multiple cognitive measures available at post medical and medical assistance.	Due to the uncertain context of addiction behaviour, the measuring and comparative values still have challenges for predictive measure design.
The familiar addiction context discussion of the root causes and consequences [17].	The knowledge of analysing root causes provides internal alertness and motive to build cognitive model development.	Uncertain addiction behaviour and different ways of human lifestyle have the challenges to make static predictive model development.

8.3 Cognitive Intelligence Role on Addiction Prediction

Cognitive computing intelligence is an essential to analysing human behaviour. Since most victimless crimes affect psychological and physical health, for psychological counselling and treatment, diagnosis of the cognitive disorder. Cognitive intelligence offers the facility of multi-facetted

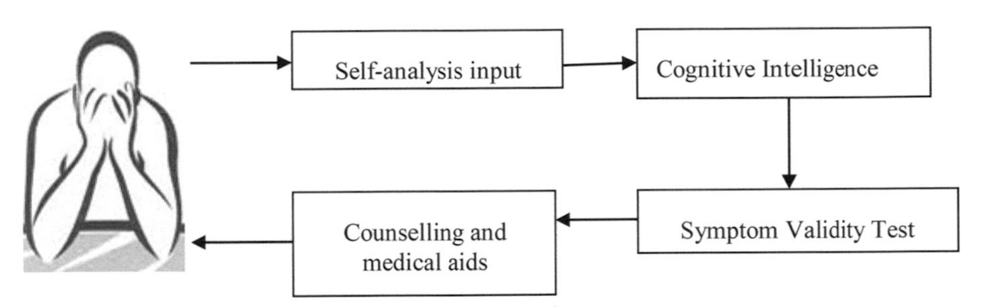

Figure 8.1 Cognitive intelligence role in victimless crimes.

technology. It provides the whole outcome as unique and accurate, as shown in Figure 8.1.

Multiple virtual based addictive behaviours cause self-destructive disorders. This self-destructive behaviour primarily affects cognitive systems and leads to mental illnesses such as depression and nomophobia. However, this type of addiction-related disorder allows for multiple direct and indirect assaults on self-destructive characteristics. Cognitive intelligence provides a classification of stages of self-destructive behaviour based on self-analysis of the individual report. This needs to be addressed as it assists in predicting appropriate analysis.

8.3.1 Self-analysis input

Dynamic contexts and multiform addiction behaviours cause different traits in actors. Since the variety of consumption and intensity of use of virtual applications are dynamic in the actor's behaviour, there are essential requirements for analysis of their volume of performing similar actions. Based on the input accuracy, the appropriate outcome, treatment, and predictive measures can be attained effectively.

8.3.2 Cognitive intelligence

Cognitive intelligence provides practical analysis and decision making on human self-destructive behaviour. Based on the input of a self-analysis report, cognitive intelligence can predict an individual's destructive level since cognitive intelligence can analyse the cognitive-based disorders and help to make a person self-aware of their destructive behaviour. Moreover, many virtual and victimless crimes can be addressed through cognitive intelligence and support predicting self-destructive behaviour [24].

8.3.3 Symptom validity test

A symptom validity test identifies the danger zone or normal zone behaviour. Furthermore, many more victimless crimes need to be analysed to diagnose the symptoms of destructive behaviour. A symptom validity test effectively aids in preventing the destructive behaviour. However, symptoms are used to identify the level of destructive behaviour at the initial stages. Cognitive intelligence systems can identify this. Based on symptoms, symptom validity tests can be tracked for either substance or related behavioural disorders [25].

8.3.4 Counselling and medical aids

Cognitive intelligence and symptom tests can aid in preventing the destructive behaviour. However, self-destructive behaviour can be helped through counselling and medical aids before reaching critical addiction. Cognitive analysis and treatment requires multiple social partners such as parents, teachers, and medical treatments. Furthermore, we can analyse the destructive behaviour and cognitive destruction based on self-harm as a victimless crime and a consequence of a cognitive disorder.

8.3.5 Self-destructive behaviour

There are multiple causative factors, and addiction formation is illustrated in Figure 8.2. The dangerous parts of virtual addictions lead to loss of physical and mental health. Moreover, suicide attempts have been recorded due to multiple gaming and gambling addiction behaviours. Hence, we discuss addiction contexts such as gaming, gambling, social media, and e-commerce based on self-destructive behaviours.

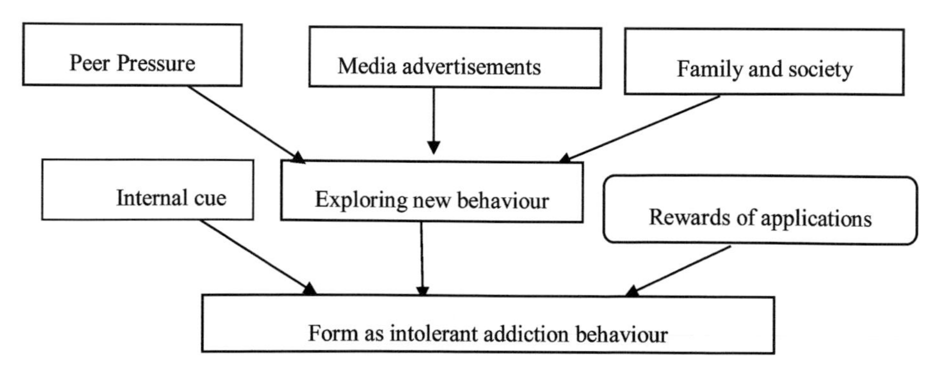

Figure 8.2 Causes and formation of self-destructive behaviour.

8.4 Causative Factors of Self-destructive Behaviour

Multiple causes influence addiction behaviour directly and indirectly. Numerous previous studies have revealed that there are multiple factors inducing addiction behaviour. We have found that significant influencing factors among those stimulus factors are

- peer pressure
- media advertisements
- society and family.

8.4.1 Peer pressure

Multiple factors stimulate substance and behavioural related addiction. When a child is a toddler, companions as peers have a vital role in influencing behaviour. Exploring new experiences and motives to perform the activities are induced by peer pressure. Due to the fun factor and curiosity to explore new activities, peers have significant influence in stimulating the initial addiction behaviours.

8.4.2 Media advertisements

The media have a high potential to make cognitive supplements via advertisements and other influences. In recent decades, multiple stimuli advertisements have been cognitively inducing access to and making addiction and dependent behaviour. For example, advertisements and marketing supplements enhance e-commerce addiction to continuously purchase, even unnecessary.

8.4.3 Society and family

Society and family has one of the most effective addiction inducing factors for addiction behaviour. Society is a place of multi-context and people joining together such as schools, malls, and public places. Hence there are possibilities to stimulate the numerous behaviours towards cognitively motivating and exploring new behaviours due to the behaviour of family members also having the potential to induce addiction behaviour.

A person explores novel experiences, and these give cognitive rewards to the person. Cognitive rewards lead to virtual addictions. Figure 8.3 depicts the concern factors of virtual addiction.

Figure 8.3 Concern factors of an addiction.

8.4.4 Context for consistent addiction behaviour

Due to multiple stimulus factors, initial behaviour is formed as exploration of experience activities. After exploring the experience, two vital aspects need to be analysed: the person is constantly connecting and regularly making use of or inducing repeated consumption and performing similar behaviour. The person is continuously addicted to using digital devices because of the reward of the applications and new experiences in the digital devices.

8.4.5 Internal signals

Cognitive stimulus is generated by neurological circuits such as dopamine and brain activity based neurons forming. For every action, the different forms of stimulus signal are generated based on positive and negative impulsive functions consistent with those repeated actions due to internal intrinsic signal as brain signal transmitters are triggers that cognitively stimulate the addiction behaviour.

8.4.6 Cognitive rewards

An internal signal consistently triggers an addiction behaviour. Cognitive rewards and internal signal simultaneously connect the context to perform similar types of activities repeatedly. Cognitive rewards such as points in-game offered in e-commerce discount sales on food and services, internally create the curiosity to stimulate access and perform similar activities. Furthermore, we can discuss the seriousness of virtual addiction based crimes.

8.5 Victimless Addiction Crimes

Many victimless crimes have been identified as well as smartphone and virtual addiction behaviour, such as suicide attempts, sexual harassment stimuli, and financial and physical health loss. The victimless crimes directly affect

Figure 8.4 Samples of virtual addictions.

those actors and indirectly affect their society. Multiple self-destructive behaviours have been identified via virtual platform based addictions, as shown in Figure 8.4.

Due to smartphone use, multiple inducing factors make users addictive through various applications, either entertainment applications or provisional services. Hence we discuss four important platforms: gaming, gambling, social media and e-commerce applications and their crime based consequences.

8.5.1 Gaming addiction

Multiple sources, directly and indirectly, encourage gaming exploration such as peer pressure, media advertisements, society and family. In recent decades, multiple virtual gaming applications have made people and their families play together through online group gaming applications. Due to the rewards of virtual application points, most gaming players regularly spend time in virtual mode. The mind stimulates the signal and regularly motivates accessing those applications without a second thought. However, gaming is entertainment that becomes a dangerous weapon while playing, spending a lot of time on it in extreme cases.

8.5.2 Suicidal attempt crimes

Multiple virtual gaming applications cause many emotional attacks such as insecurities in their lives and trigger them to spend much more attention and money to use those applications. Due to the attractive visual and virtual design of gaming applications, users regularly use those applications and motives to play the next level of gaming without a second thought. For example, many suicidal attempts have been recorded while playing blue whale gaming applications in recent decades. Similar to suicidal attempts, multiple self-harm is cognitive, such as depression anxiety and decrease in self-confidence.

8.5.3 Social media addiction

There are numerous forums available for social communication in the digital decade, such as Facebook, Instagram, Twitter, etc. The problem of these forums using multi-feature facilities is that they make the user repeatedly notification check and regularly use these applications. For example, regularly uploading videos and photos on Facebook, exploring things in social media and often checking social media to check the subscriber and viewing of sharing content.

This makes users focus on the attention-seeking of others, and encourages poor oral communication and lack of real-world social interactions. Since virtual addiction builds up on social media, seeking others to provide recognition leads to multiple direct and indirect negative mental illnesses such as unipolar and bipolar depression.

8.5.4 Gambling addiction

The increasing volume of digital games has resulted in inevitable consequences [26]. Multiple gaming applications lead to a path of gambling addiction, as entertainment provides a facility to play after paying, and users can virtually lose those games. Gambling applications indirectly lead to financial loss to the player, and rewards and internal cues are stimulated through initially offering small acknowledgement and finally making them into financial traits as victimless crimes.

Gambling addiction causes the player to lose awareness. Mental illness, decision making, and complex and rare processes of self-analysing lead players to risk. However, mental illness also affects physical health. High financial traits lead the player to lose wealth, harm to self, and indirectly affect others. Online gaming addiction (OGA) is already rising, with severe ramifications, including neglecting family obligations, job loss, health complications, criminality, and perhaps even death [27-28].

8.5.5 E-commerce addiction

Multiple e-marketing sites tend to designed to be attractive and make customers regularly purchase their products. This e-commerce application makes their customers by notifying the products on display at their display screen. The rewards offer discounts cognitively stimulate the customer for making regular customers.

E-commerce addiction can make the buyer regularly purchase as an addiction leads to financial losses. E-commerce addiction makes the user buy compulsively and even unnecessary things. Every product, food and service makes the user more dependent. These models are more sophisticated but, at the same time, leads to a lot of financial and mental losses in extreme addiction cases.

8.6. Virtual Addiction Crimes

Due to addiction, interpersonal and intrapersonal skills are destroyed, as the addiction leads to forgetting or hiding other aspects of life such as education, relationship, and self-discipline. Extreme levels of addiction cause lack of awareness and reduces the time for self-analysis and developing internal signal to perform normal activities.

Figure 8.5 illustrates the consequences of victimless crime based addictive behaviour. Negative thoughts, suicidal thoughts, low self-esteem, sleeplessness, decreasing focusing and learning abilities, and loss of physical and mental health stabilities are induced.

These cognitive imbalance states create the context for abusing and attacking the actor directly and indirectly. Virtual addiction can induce cognitive compulsive and craving tendencies based on the volume of consumption behaviour. However, before entering the danger zone, it is essential to take measures of predicttion through cognitive computing-based applications. Hence an artificial intelligence-based cognitive system has the potential to accurately predict results effectively. Moreover, many virtual activities stimulate multiple self-harm-based disorders, which are likely to be victimless crimes. The forensic investigation team faces challenges in finding the virtual crime because most virtual crimes are victimless. However, any effects such as self-harm and harm to others are treated as a crime. In this context,

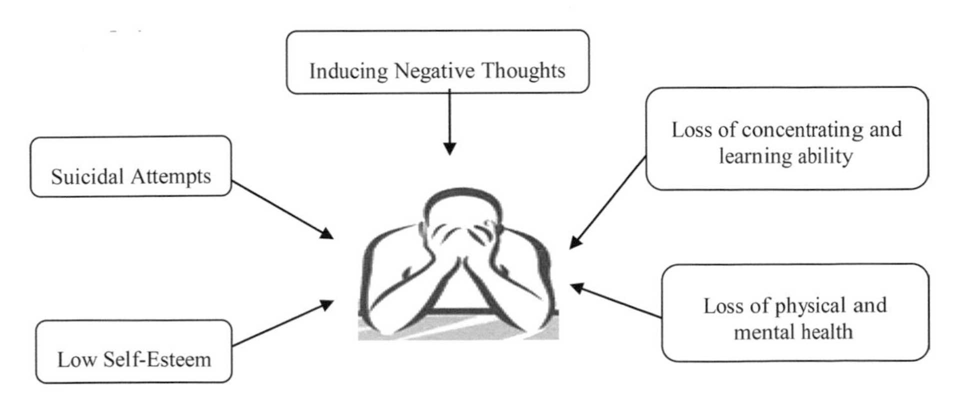

Figure 8.5 Impacts of victimless self-destructive addiction behaviour.

our study work has evolved, and cognitive intelligence with victimless crime identification has become essential to make the digital and virtual world aware of victimless crimes.

8.7 Limitations and Future Directions

Several kinds of literature are discussed that exhibit physical healthcare systems and implementations. However, several investigations are still needed into the emotional and physiological cognitive care processing fields. Our research focused on cognitive psychological victimless crimes, which are extremely important to consider in today's digital system evaluations. Our research will continue to invest in developing cognitive models for psychological healthcare management and the prevention of self-destructive behaviour based on mental and psychological attentiveness.

Study and natural intelligence interpretations and psychological connection in psychological stability should be used as pretreatment assessments for cognitive impairment. However, psychological self-measure strategies for self-analysis in cognitive addicted behaviour patterns are desperately needed.

8.8 Conclusion

Virtual addiction based self-destructive behaviour leads to multiple victimless crimes. Cognitive intelligence provides a model used to validate the symptoms for preventing the addiction danger zone. This analytical model becomes more beneficial to the actor for addiction states of self-analysis. Cognitive healthcare and intelligence systems efficiently support addiction prediction at a high level. We examined addiction-based victimless crime and the importance of cognitive intelligence in developing models of symptomatic validity tests as addiction aware systems in this chapter. Overall, this study assists in studying the stimulus and cognitive intelligence on cognitive addiction based victimless crimes.

Furthermore, we intend to be aware of digital-based victimless crimes. It is essential to make cognitive intelligence in cybercrime investigation assist models to prevent virtual-based victimless crimes. AI and an integrated technology model will be employed in the future to develop psychological attention for addiction prevention strategies.

The widespread use of computing technology through smartphones navigates the user into the virtual world. Multiple features of virtual platforms have the potential for the user to access those technologies continuously. However, these gadgets make users more dependent on those applications and services. Furthermore, increasing smartphone use is the root of multiple cognitive affecting disorders, which induces self-harm and also indirectly harms others as a consequence; virtual cognitive addiction leads to acting multiple traits while the user is at the addiction stage. The addiction becomes self-destructive behaviour.

8.9 Acknowledgements

Our sincere thanks to the Department of Computer Science and Engineering and the management of SRMIST for motivating us.

References

[1] M.S. Fryman & R. William, 'Measuring smartphone dependency and exploration of consequences and comorbidities',Com.in Hum.Behav. Rep, 100–108, 2021.

[2] R.M Ryan, et al.,' The motivational pull of video games: A self determination theory approach', Motiv. and emo, 30(4), 344-360, 2006.

[3] D.Tran,'Classifying nomophobia as smart- phone addiction disorder', UC Merced Under. Res. Jou, 9(1), 2016.

[4] C.Yildirim,'Exploring the dimensions of nomophobia: Developing and validating a questionnaire using mixed methods research', (Doctoral dissertation, Iowa State University), 2014.

[5] A.L.S King, et al.,'Nomophobia: Dependency on virtual environments or social phobia?',Comp.in Hum.behav,29(1), 140–144, 2013.

[6] S. Uysal,et al.,'Social phobia in higher education:the influence of nomophobia on social phobia',The Glo. e-lear. Joul,5(2), 1–8, 2016.

[7] P. Gilbert,'The relationship of shame, social anxiety and depression: The role of the evaluation of social rank'.Clinical Psychology & Psychotherapy: An Int. Jou.of Theo. & Prac,7(3), 174–189, 2000.

[8] D. Gentile,'Pathological video-game use among youth ages 8 to 18: A National study'. Psycho. sci, 20(5), 594–602, 2009.

[9] C.A Anderson, D. A., Gentile, & K.E Buckley,'Violent video game effects on children and adolescents: Theory,research, and public policy' Oxford University Press, 2007.

[10] Z. S Chen, Z. S., & M.C Chung, (2016). The relationship between gender, posttraumatic stress disorder from past trauma, alexithymia and psychiatric Comorbidity in Chinese adolescents: A moderated mediational analysis. Psy.Qly, 87(4), 689–701, 2016.

[11] X. Liu, & G.Yoo, Relationship between Chinese adolescents' academic performance and smartphone overdependence: Moderating effects of parental involvement. of Fam. Rels, 22(4), 157–179, 2018.
[12] S I Chiu, The relationship between life stress and smartphone addiction on taiwanese university student: A mediation model of learning self-efficacy and social self-efficacy. Comp. in Hum.Behav, 34,49–57, 2014.
[13] N.S. Hawi, &M. Samaha, Relationships among smartphone addiction, anxiety, and family relations. Behaviour & Information Technology, 36(10), 1046–1052, 2017.
[14] S.-M. Chang, G.M. Hsieh, S.S. Lin, The mediation effects of gaming motives between game involvement and problematic internet use: escapism, advancement and socializing, Comput. Educ. 122, 43–53, 2018.
[15] G.J. Hyun, D.H. Han, Y.S. Lee, K.D. Kang, S.K. Yoo, U.S. Chung, P.F. Renshaw, Risk factors associated with online game addiction: A hierarchical model, Comput. Hum.Behav. 48, 706–713, 2015.
[16] S. Sarker, M. Ahuja, S. Sarker, Work–life conflict of globally distributed software development personnel: an empirical investigation using border theory, Inf. Syst. Res. 29 (1), 103–126, 2018.
[17] G.J. Hyun, D.H. Han, Y.S. Lee, K.D. Kang, S.K. Yoo, U.S. Chung, P.F. Renshaw, Risk factors associated with online game addiction: A hierarchical model, Comput. Hum.Behav. 48, 706–713, 2015.
[18] J. Resch, J. Ehrentraut & M. Barnett-Cowan, Gamified,'Automation in Immersive Media for Education and Research'. arXiv preprint arXiv:1901.00729, 2018.
[19] S.W Choi,et al., 'Comparison of risk and protective factors associated with smartphone addiction and Internet addiction',J.of behave. addict, 4(4) 308–314, 2015.
[20] S. Haug, et al., 'Smartphone use and smartphone addiction among young people in Switzerland'. J. of beha.addict, 4(4), 299–307, 2015.
[21] S. Aker, et al.,' Psychosocial factors affecting smartphone addiction in university students'. J. of Addict.Nursing, 28(4), 215–219, 2017.
[22] N.S. Hawi, M. Samaha, 'The relations among social media addiction, self- esteem, and life satisfaction in university students'. Soc. Sci. Com. Rev, 35(5), 576–586, 2017.
[23] K. Demirci et al., A, 'Relationship of smartphone use severity with sleep quality, depression, and anxiety in university students'. J. of behav, 4(2), 85–92, 2015.
[24] V. Sabapathi & K.P. Vijayakumar, 'A Study of Addiction Behavior for Smart Psychological Health Care System'. Role of Edge Analytics

in Sustainable Smart City Development: Challenges and Solutions, pp. 257–27, Wiley Online Library, 2020.

[25] Block Steven, 'Victimless Crime', 75-79, 2015.

[26] Zhang, X., et al., 'Artificial intelligence in cognitive psychology—Influence of literature based on artificial intelligence on children's mental disorders', *Agg. and Viol. Behav*, 101590, 2021.

[27] Hsu, Shang Hwa, Ming-Hui Wen, and Muh-Cherng Wu. "Exploring user experiences as predictors of MMORPG addiction." Comp. & Edu, 53(3), 990–999, 2009.

[28] Turel, Ofir, Christian Matt, Manuel Trenz, Christy MK Cheung, John D'Arcy, Hamed Qahri-Saremi, and Monideepa Tarafdar. "Panel report: The dark side of the digitization of the individual." Internet Research (2019).

9

The Future of Artificial Intelligence in Digital Forensics: A Revolutionary Approach

Ishi Saxena, G. Usha, N. A. S. Vinoth, S. Veena, and Maria Nancy

Department of Computing Technology, SRMIST, KTR
Email: ishisaxena3@gmail.com; ushag@srmist.edu.in;
vinoth.nas@gmail.com; 4veena.s@gmail.com; marianancyrajg@gmail.com

Abstract

Digital forensics (DF) is an area that is turning out to be progressively significant in collecting a lot of computationally large and complex information and frequently requires intelligent analysis throughout the globe. DF presents a method for carefully utilizing scientific information to reveal and understand electronic evidence. Artificial intelligence (AI) techniques address challenges in digital forensics and give a method for handling computationally enormous or complex issues in a reasonable period. This chapter provides an overview of some of the AI principles and procedures. It features everyday difficulties, including the accessibility of data sets in a few regions and explicit trouble in clarifying the outcomes when specific strategies are utilized. This chapter suggests that there is indeed a requirement to improve the utilization of the AI principles accessible and also discusses the strengths and constraints of currently used investigative technologies. This chapter also suggests the difficulties in delivering models where limited training information is induced from the models. Therefore, in this chapter, we have proposed a taxonomical framework with graphs and results to battle the current and analyze the future difficulties of DF.

9.1 Introduction

Digital forensic science includes the recovery of proof from advanced gadgets. It is in some cases characterized as cycle models that catch the phases

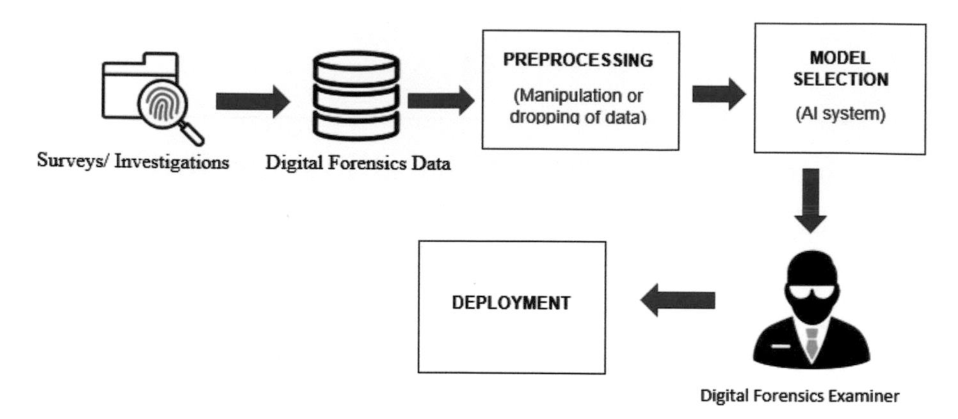

Figure 9.1 Future of artificial intelligence in DF.

of an investigation. In the explanation of this section, the cycle is split into stages that help communicate where AI methods have been applied to an advanced investigation. These are acquisition, examination, analysis, and presentation. A genuinely explainable artificial intelligence system develops a type of thinking that involves human aspects of the incoming data to describe the model's judgement process. To be trusted, these platforms must prioritize the demands of end-users (for example, several user groups within a given sector or various parties within the legal system). Creators of these technologies should collaborate with the people who will use them, or they risk organizational work that is too opaque or difficult to be understandable.

The future of artificial intelligence in digital forensics is depicted in Figure 9.1.

Recent advancements in artificial intelligence (AI) have resulted in a rapid increase in such approaches across many businesses and technology. The increasing complexity of AI methods makes assurance and testing of such systems' safety and dependability problematic. As a result, the growing area of AI safety attempts to address the AI-enabled system's dependability and safety [8]. Digital forensics aims to create the tools, methodologies, and procedures needed to conduct a forensic investigation of AI failures. In addition to examining underlying technological flaws, such investigations must evaluate if the breakdown was triggered by intentional behaviour and who is responsible for the damage. Technology, legislation, criminology, psychology, and other fields are all involved in this subject [9].

Machine learning (ML) has been frequently used in digital forensic investigations for data discovery, device triage, and network forensics. Task specification, feature creation, and assessment and optimization are the three phases in ML applications. Depending on the kind of target labels, an ML task can be classified as either a classification/clustering or a regression challenge. Feature transformations and selections are made during the experiment to reduce over-fitting, enhance performance, and shorten training time. In the same way that humans do not develop features, the digital forensic applications of deep learning (DL) are analogous to machine learning (ML). Instead, a general-purpose learning technique is used to learn from data. Optimization and inference are the two steps of a DL model. The values connecting the units called neurons specified in the model are changed via the training phase. Interpretation is used to make accurate predictions based on unlabeled data not visible during learning [10].

As per recent trends, the setting for crime and the application of digital forensics is developing and extending in scope. This same advent of new technologies and surroundings for prospective cybercrime, such as breakthroughs in high computing and the cloud, and the spread of social media and communications devices, requires a review of the tools and methodologies used by digital forensics investigators.

More efficient and appropriate practices and techniques in digital investigations must be investigated [9]. We propose that intelligent digital forensics methods be used to speed up and increase the efficiency of the inquiries. This study identified artificial intelligence tools and theories of digital forensics intelligence and intelligent investigations as a potential possibility [11].

It is critical to understand how AI is used while investigating AI as an instrument of crime. Authorities should gather and interpret datasets, learning methods, a variety of skills, inference models, and implementation of the AI system used to break the law. Investigators should understand the AI's goal is based on the assessment. Investigators must be able to distinguish between the developer's purpose and the AI output in this scenario. AI programs, unlike traditional programming, frequently have unexpected outcomes [12].

In traditional programming, data and code are processed to produce a result; in AI, information and production are used to create a program. AI parameters are typically set with some randomness since many AI models use arbitrary values in the training stage. As a result, algorithms with different parameters and outcomes can be built using the same information and training model. It would be impossible to demonstrate if AI has been used as a weapon, how AI was deployed, and what harm AI caused because investigators would be unable to recreate the case [11].

Another roadblock in functional AI investigations is complexity. An artificial intelligence model employs a variety of methods and modules in general. Several AI designs incorporate application programming interfaces

(APIs) provided by modern AI systems for efficiency. Due to the AI system's complexity, the investigators must have a thorough understanding of AI and the ability to reverse engineer it [12].

Just a portion of the AI system that may be gathered confounds forensic investigators. Collecting all elements, such as the development system, cognitive approach, information, trained model, and supervised learning, is nearly impossible due to technological and regulatory challenges. Because criminals try to delete evidence of their crimes, it is necessary to establish the design and trace of AI using little data. Due to the scarcity of evidence, investigators will have difficulty replicating the AI system under investigation. This is a big deal for investigators since repeatability is one of the essential concepts in digital forensics. Throughout this chapter, we state that to address existing and future cybercrime challenges, it is essential to increase the use of existing resources and move further than the skills and limitations of the current automated analysis.

9.2 Artificial Intelligence in Digital Forensics

Artificial intelligence in digital forensics is classified in Figure 9.2.

We see AI forensics as a subfield of digital forensics, defined as the scientific and regulatory tools, techniques, and proccdures for collecting, gathering, analysing, and disclosing digital material related to AI-enabled breakdowns [12]. The rise in cyberattacks and the complexity of the types of cybercrime, combined with time and budget constraints, both supercomputing and human, in attempting to address cyberattacks, has put a massive strain on digital forensics experts' opportunity to implement forensic analysis

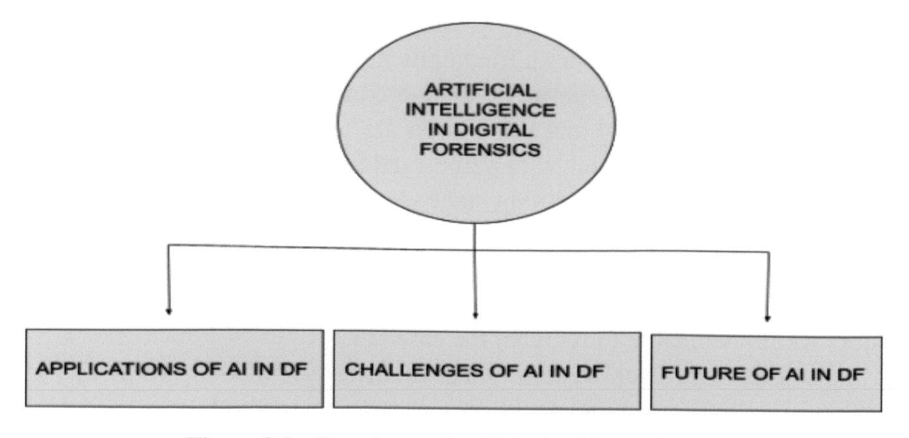

Figure 9.2 Key phases of artificial intelligence in DF.

Table 9.1 The difference between future and applications of AI in DF.

Applications of AI in DF	Future of AI in DF
If any information survives, documents erased inside one file system might be retrieved deterministically. Nevertheless, this information is missing in certain situations, and the material is stored in the piece's unclaimed areas.	ML approaches can speed up the acquisition of essential data for inquiries by using the knowledge gained through earlier computer forensic assessments.
Network inquiries are frequently part of a broader inquiry that includes incident response, mobile devices, smart-watches, financial fraud etc. Several gadgets or technology that have already been connected are usually involved in such inquiries.	Network activity monitoring will become more tiered, allowing for improved matching of user information across many systems or networks. Single suspicious consumers and their behaviours may now be profiled more broadly using contemporary networks and gadgets.
DL approaches emphasize significant moments in a timeframe depending on positive or negative emotion in message representations of events, such as "lost login" or "verification loss."	A time frame representation of the action is only one perspective of an information gathering; it may be a helpful access point into a database since it allows the perspective of being "pivoted" to a document suitable to analyze, similar material, and again to timeframes.

and digital investigations process for obtaining immediate results. The difference between future and applications of AI in DF is discussed in Table 9.1.

9.2.1 Applications of AI in DF

AI, or artificial intelligence, studies intelligent agents or agents that react to their environment to find the best path to their objective. Machine learning (ML) and deep learning (DL) are the two main disciplines of AI in computer science [12].

AI's success is data-driven because no clear coding dictates a particular result.

Digital intelligence can impact advanced legal sciences in furnishing aptitude to assist with the normalization of the portrayal of information and data in the computerized scientific space. The data available when constructing artificial intelligence systems in digital forensics are outlined. The datasets used to prepare the models are simple, and data pre-handling is an important stage in machine learning.

AI is a vast discipline with approaches for solving a wide range of issues in various fields. AI tries to mimic human intelligence while also containing the subfield of machine learning, which is the foundation for intelligence development. It can investigate information without the requirement for human contact and foster more superior knowledge with every reiteration because of calculations [13]. AI is the most common way of programming PCs to boost an exhibition basis by utilizing model information or earlier information.

Deep learning is a branch of machine learning research that solves issues by analyzing big datasets and employing neural networks. According to Jackson (2019), a neural network is a very simplified representation of a biological brain network, portrayed as a collection of interconnected "artificial neurons." These networks make judgments based on data input and modify values depending on feedback to come closer to the intended output [14].

Because of its capacity to adapt without requiring user input, artificial intelligence is beneficial in various situations. Many solutions have recommended using machine learning to deal with the massive volumes of data that digital forensics requires.

As previously stated, one approach that employs machine learning techniques is k-means clustering. k-means clustering is defined by Kapil et al. (2016) [15] as an "unsupervised techniquc uscd to clique distinct objects into groups." A04 offers a map reduce-based k-means clustering method for big data analytics that can successfully manage massive volumes of data. A basic description of map reducing comprises two phases: a mapping phase that groups similar things together and a reduce phase that summarizes the objects. When dealing with vast quantities of data, A04 integrates this model with the k-means clustering method to improve efficiency.[15]

A29, in a similar vein, recommends using map reduction in conjunction with k-means clustering to enhance efficiency on big datasets. These algorithms improve performance while ignoring privacy concerns. Another benefit of this solution is that it may be outsourced to cloud servers, lowering the cost of in-house IT systems.

Neural networks are a subset of machine learning that creates a network of neurons and synapses that operates similarly to the human brain [16]. They are the foundation for the idea of deep learning, which was previously discussed. In significant data contexts, A12 is experimenting with neural network approaches to deal with network anomalies [17]. A25 discusses using neural networks to manage large amounts of data and gather evidence. A10 uses deep learning in cyber forensics to combat the rising number of cyber-attacks aimed at both individuals and businesses.

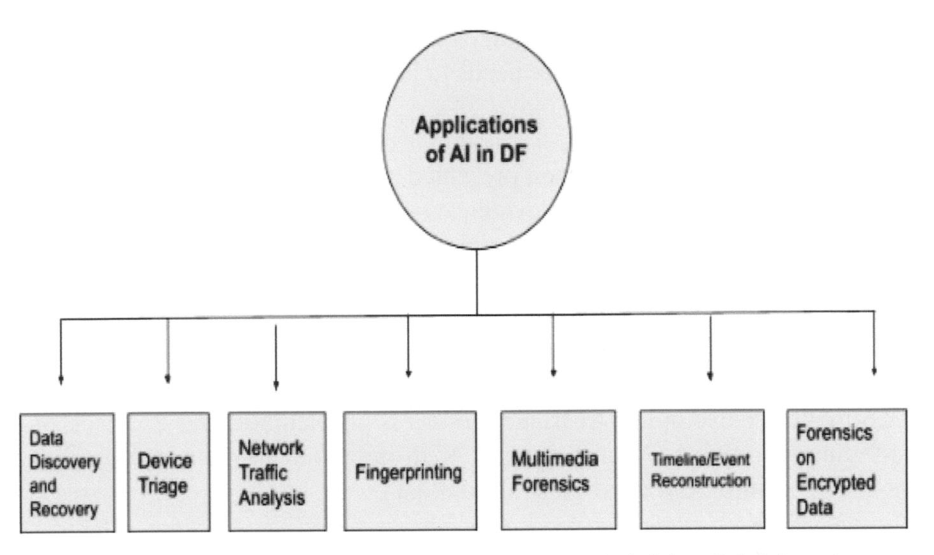

Figure 9.3 Seven ways in which artificial intelligence in helping digital forensics.

Deep learning may also assist in discovering not safe for work (NSFW) pictures, as A15 discusses [18]. This method uses deep learning algorithms to reorganize large datasets of photos based on their chance of having pornographic material, allowing investigators to identify relevant images early. In contrast, the percentage of relevant images is low. This is important since most child pornography investigations involve human examination, and physically examining hundreds of photographs would take a long time. More clearly, we will discuss the classification of applications in digital forensics using Figure 9.3.

9.2.1.1 Data discovery and recovery

If any metadata survives after a file is wiped inside a record framework, it may be algorithmically recovered. In some cases, this metadata is absent, and the file is stored inside the volume's unclaimed regions. However, such records could be broken and partly rewritten when they move across the disc. The technique of retrieving such files without information is known as file carving.

Machine learning may be used to automate digital forensics investigations [8]. ML methods can speed up the acquisition of relevant information for investigations by utilizing the knowledge gained from prior digital evidence analysis. One of the difficulties in developing AI models is avoiding adversarial attacks. The attacker can interfere with the source, resulting in erroneous results. During an inquiry, any pre-trained model might lose its efficacy. This is a challenge that needs to be overcome.

9.2.1.2 Device triage

Digital evidence triage has been presented as a method for identifying, analyzing, and interpreting digital evidence on time. In the future, the effectiveness of device triage may become more essential in digital investigations. Currently, the investigating officer determines the order in which devices are acquired and processed at a crime scene. As more AI-based techniques are developed, the on-scene preliminary inspection may be able to quickly concentrate the analysis on the equipment most likely to have case-progressing information. Developing AI triage models is problematic due to a lack of a sufficiently large and shared dataset. With improved triage accuracy, fewer resources would be wasted on irrelevant data.

9.2.1.3 Analyze traffic on a network

In many cases, system inquiries become part of a broader investigation that encompasses incident response, cloud, information technology, smartphones, wearable technologies, and financial crime. After the occurrence, investigators have access to a wealth of literature on AI in data transmission offsite in the batch process [11]. Feature selection procedures significantly impact the models generated by AI technologies. Network data has been obtained, and features extracted before using clustering techniques to develop IOC guidelines for detecting attacks. Indeed, with encoded networks, artificial intelligence methods can, in any case, be utilized to demonstrate factual data of an organization.

9.2.1.4 Encrypted information forensics

Encrypted data is one of the most pressing concerns for digital forensics investigators worldwide. Cryptographically secured devices and data ubiquity make digital forensic inquiry impossible [20]. The forensic disc image becomes useless if the device under examination employs disc encryption. Gaining valuable information without having access to the encoded information and executing cryptographic fundamental retrieval assaults are two possible paths for electromagnetic and power side-channel analysis that can be aided through AI methods. Using electromagnetic and power side-channel observational data, several AI methods have been applied to the first aim.

9.2.1.5 Event restoration

In digital forensics, event restoration is a two-step procedure. The first is simply recognising that something happened, or more precisely, that something happened at a particular moment. The second entails examining timestamps that can be recovered from digital forensic evidence [12]. Attempts have been made to analyze this large number of timestamped data using automated methods. Neural organizations are suitable for managing vast volumes of information due to their parallelism and speculation capacities. Studiawan utilized DL strategies to feature revenue occasions in a timetable dependent on certain or negative feelings in the message-based portrayal of occasions.

9.2.1.6 Forensics of multimedia

Audio, video and image evidence are investigated in multimedia investigations, a subset of computer forensics. It does not have to be restricted to smartphones; surveillance analysis may also be included. Europol has utilized a social media crowdsourcing technique to track down evidence in the fight against child abuse. In digital forensics, the advent of CNNs has permitted astonishing and promising findings [21]. Automated object recognition in low-resolution pictures is necessary. Images of low visibility and that are challenging to the human eye can be managed using customized systems trained on low data quality. There is an apparent demand for shared, well-curated datasets in the scientific community.

9.2.1.7 Fingerprinting

Digital forensics is seeing a surge in device fingerprinting [22]. It may be used to determine the model of the source camera and the individual camera. The job of fingerprinting fits well with AI categorizations. For instance, malware classification has been a popular software topic with many past studies. Knowing the camera type employed to record a video might be very helpful in this case. Device and user behavioural fingerprinting may significantly assist anomaly detection. This might lead to more effective host-based and network-based malware detection for devices connected.

9.2.2 Challenges of AI in DF

A variety of current difficulties in digital forensics are explored in this chapter. Each of these issues can make it difficult for digital investigators and detectives to find relevant information in a variety of situations needing digital forensic investigation. The negative impact of these problems can be substantially magnified when they are combined. Due to these difficulties, as well as a lack of competence and large workloads, many law enforcement organizations worldwide have a digital evidence backlog in the order of years.

One of the most significant issues with using machine learning in digital forensics is the lack of certainty in defining artificial intelligence method usage in the thinking process [23]. This can be addressed by seeing the application of artificial intelligence in digital forensics as mainly focused on outlier detection. There are two parts to this: constitutional and mathematical. Acts that violate a jurisdiction's law, such as underage drinking or driving, are known as legal oddities.

AI-based technology can only be used to help in investigations, which human investigators must still oversee. To some part, the forensic outcome's correctness is determined by the human investigator's ability. The investigator must either receive extensive training and skill development or hire highly competent investigators to overcome this obstacle. The growing volume of data and information collected and subjected to digital forensic investigation is a challenge that new and better AI technologies must address. Steganography, encryption, and anti-forensics media formats, for example, might be used. Figure 9.4 explains the challenges of AI in DF.

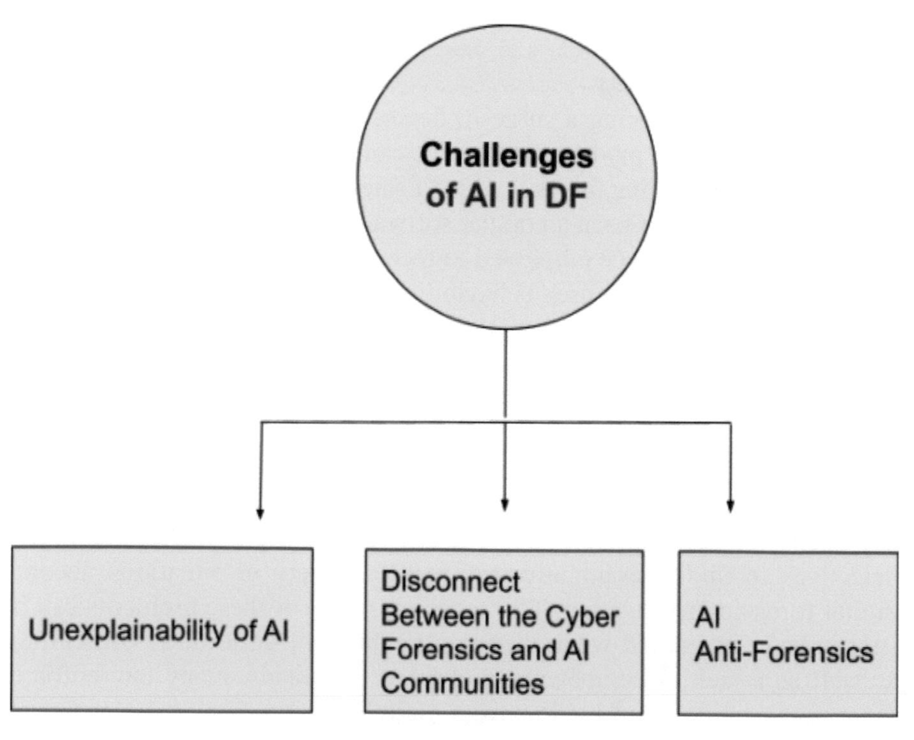

Figure 9.4 Key things to assess before starting your artificial intelligence journey.

9.2.2.1 Unexplainability of AI

It may be necessary to provide clear and precise analyses of the decision-making procedure that led to the unwanted practices in order to conduct a thorough and undisputed underlying cause investigation of AI failures. Be that as it may, the examination of the logic of complex simulated intelligence frameworks is still in its initial phase, and the cutting edge is a long way from taking care of the issue of logic. Moreover, some new writers (e.g., Yampolskiy, 2019) contend that as artificial intelligence innovation and abilities advance over the long haul, it may turn out to be more troublesome or even inconceivable for AI frameworks to be logical [24]. In such conditions, more direct reflections of the dynamic interaction might upgrade the scientific investigation of such disappointments. For example, Behzadan et al. (2018) [24] propose a psychopathological deliberation for complex AI security issues. Comparable reflections might be needed to empower the detailed scientific investigation of progressed AI.

9.2.2.2 AI Anti-forensics

In digital forensics, criminals are continuously adapting to technological advancements and employing tactics such as decoys, fake evidence, and forensic cleaning to thwart forensic investigations. Anti-forensics tactics like this are expected to be created and used by criminals to influence AI forensic investigations. As a result, proactive detection of such approaches and the development of mitigating remedies will become a more prominent topic of research in this field. Researchers recently created an anti-forensics general taxonomy [21]. However, AI anti-forensics was not included in the taxonomy. While this is a difficulty, it is also an opportunity for researchers.

9.2.2.3 Disconnect between the cyber forensics and AI communities

The gap between the AI Security and Digital Forensics groups constitutes one of our major problems. Because researchers from those two fields are not collaborating, the field of AI Forensics is yet to be defined and is open to additional research. In a recent poll research, most of digital forensic practitioners (67%) (disagreed, agreed, or were indifferent) on their expertise in Data Science, making this disparity clear (Sanchez et al. 2019) [25].

9.2.3 Future of AI for DF

There is already much work being done in applying AI to certain aspects of digital forensics. This section covers general problems and possible possibilities, such as uncharted territory where AI methods have yet to be used. A prominent priority area is improving technique precision. The absence of big, clean, annotated datasets in some domains makes it difficult to identify and train models and measure their correctness. It is an exceptional circumstance whereby an AI system updates in live time, for instance, benefiting from ongoing investigations to allow faster proof discovery in future situations.

The approach output might regularly alter in this scenario, creating a substantial confirmation difficulty. Furthermore, it is worth thinking about whether sharing models are suitable in specific situations.

DF examiners may benefit from XAI by classifying images, for instance, while maintaining their psychological safety [8]. It can look at resembling measures of a picture without looking at it, or it can read specific predetermined interpretations (generated by XAI/ML technology) of pictures by not looking at them. One example is how AI and machine learning may assist in classifying child pornography and other unpleasant pictures [26].

XAI may be of little help in supporting the investigation process on the first pass. After a suspect has been captured and evidence is assumed to exist, the bulk of digital forensics investigation is performed. Investigators frequently request that DF specialists extract data from devices and give it to the courts in an easily reportable manner [8]. The alternatives given by XAI to deal with problems in DF are described in the following sections from the perspective of law enforcement.

The preceding sections have demonstrated that substantial work in using AI in certain aspects of digital forensics already exists. This section covers general issues and possible possibilities, such as uncharted territory where existing and new AI approaches have yet to be used.

Enhancing technique reliability appears to be a focus area for overall issues. Training models and determining their accuracy in digital forensics is complex because of a lack of big, clean, labelled datasets in some domains or existing datasets that are not publicly available [27]. While many datasets may be used to train computer vision-based algorithms, delicate information like CSEM are clearly and logically restricted [28]. There are no suitable datasets for whole-disk techniques in which customer experience is thoroughly researched and identified, enabling DL algorithms to detect significant characteristics.

Despite these obstacles, there are several possibilities for improving AI applications and applying AI to new areas of digital forensics. Inference of behaviour from data collected from innovative sources, such as smart homes, IoT sensors, automobile forensics, and combinations thereof, is one of them [29]. Indeed, AI approaches might be helpful in any situation where data

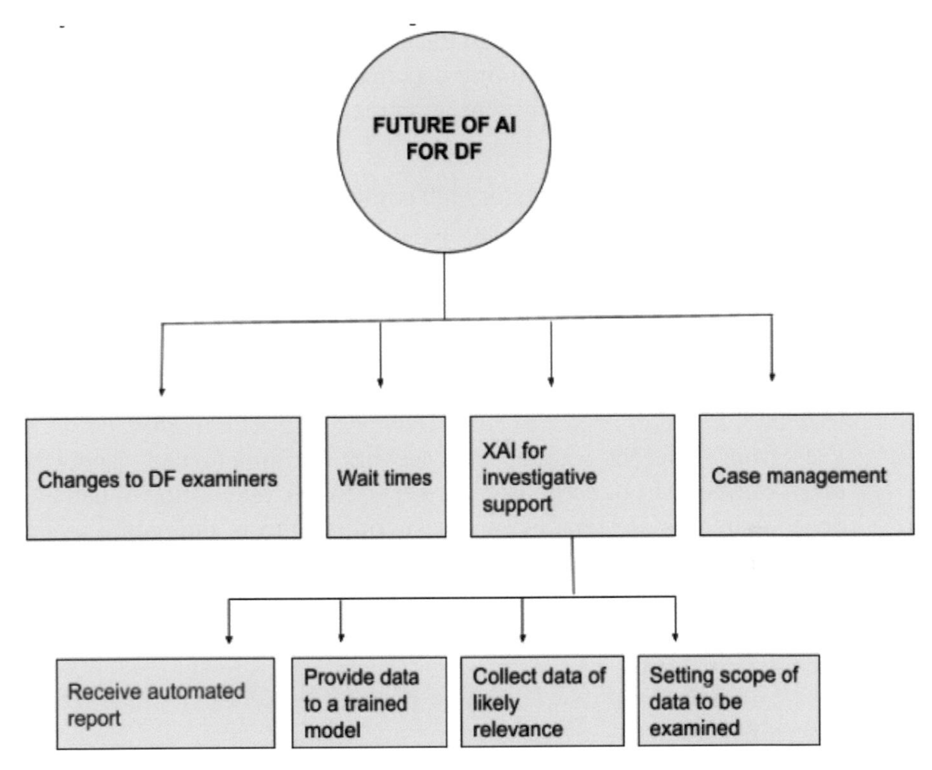

Figure 9.5 Opportunities for artificial intelligence to improve digital forensics.

from many sources is needed, such as several suspects, gadgets, or cases [30]. Figure 9.5 gives a broad idea of the future of AI for DF.

Tools and software concerning digital forensics and AI, such as EnCase, Forensic Toolkit, The Sleuth Kit etc., will be used in the future.

9.2.3.1 Changes to DF examiners

The job of a DF examiner will almost definitely alter if XAI becomes widely used in the DF sector. This would result in at least three minor modifications to the DF examiner's access privileges to trained AI data. We will go through each of these modifications one by one in the following:

1. Obtaining training data for XAI models might be tricky. In a law enforcement environment, computers would need to be educated on data similar to user information that is likely to be relevant (e.g., photographs of CAMs), which has its own set of legal constraints [1]. The justifications of XAI systems could be skewed by aspects of the

training examples that are not characteristic of the data under investigation, through features of training data that are not representative of the data under investigation, or by programmer's decisions made during the strategy interplay.

2. Supervisors and other necessary lab employees and supervisors should possess the following qualifications: the fundamental abilities, credentials, and understanding required to work in the field when AI/ML is implemented in DF. Because of the rapid pace of smartphone devices, DF examiners must also be quick learners. This problem is exacerbated by competition for the same pool of talent from other, typically higher-paying industries [2].

3. Data from different sources like network equipment that cannot be taken offline is frequently used for DF in the private sector. When evidence materials are retained under lawful seizure in the public sector, it is typically up to the investigating DF officer to handle an individual's concerns about the seizure of their property [2]. It is not feasible for proof of things to sit away for quite a long time, anticipating a DF agent's availability.

9.2.3.2 Wait times

As machines and applications become more advanced, delays and timeframes for DF operations will keep growing. Analysis may be done faster with a smaller data set, resulting in more significant influence on ongoing investigations in the short term. Some programs attempt to handle the growing data load by copying detailed data from target devices using manually preset file masks. Some systems, such as the Kroll Artifact Parser and Extractor (KAPE)[4][5], try to deal with the growing amount of data by using physically prepared record coverings. This can shorten the time to process these records, but it does not solve the volume problem. Documents retrieved using tools like KAPE must be examined by a DF investigation and announced in a meaningful order.

9.2.3.3 Management of the case

Case categorization, the difficulties of a forensic examination on particular property items, data recovery capabilities, the accessibility of DF examiner and investigator staff, and other issues have made DF case administration a complicated and unproductive operation. XAI might help DF examiners triage cases more efficiently, allowing for more efficient management of resources. XAI can also assist in determining whether DF analysis is necessary to determine a case's outcome [8]. XAI/ML can give information for preparation, characterization, and grouping. It can improve the capacity of a DF caseworker by computerizing some of them through rehashed information and actions. A DF examiner can take artefact extraction to a whole new level thanks to XAI/ability ML to comb through data and recognize context. This would save a lot of time and effort in the case of a DF case. XAI/ML is very likely to help advance effective models for predicting the location of specific things to conduct additional data analysis. Criminals, like DF investigators, use a variety of equipment and methods depending on the type of mistake they make.

9.2.3.4 XAI for assistance with investigations

XAI might eventually be used to analyze the information from a targeted system and create and include documentation with information on the relations between people, places, times/dates, and objects important to the investigation. The remaining part of this section is given to XAI for investigative support, one possible implementation of such a paradigm.

1. **Define the purpose of the data reviewed**: When conducting a DF examination, various evidence items will have varied data extraction methodologies and availability. Users would first indicate which data items should be examined [6]. This could be due to the quality of data, the significance of the suspect to the investigation, or practical competence. Although there may be a demand on DF examiners to include all seized documents in the process, as with any inquiry, the analytical goals must be emphasized.

2. **Gather data of importance**: The refining of data using AI models consumes many resources. In the same way that technologies like KAPE allow DF examiners to accelerate information gathering and begin examinations faster, they may also be used to reduce the amount of data that the XAI models have to analyze [8]. Tools must be able to get target data and prepare it for processing by the XAI system. As with any forensic investigation, data processing should be done by applicable regulations. Data gathering should be done with the proper permission.

3. **Supply data to a trained prototype**: The XAI model must be able to discriminate between various forms of data and meaningfully interpret it (e.g., the EXIF data located within a JPEG file should tell the model something different to a shortcut file) [8]. People, locations, times/dates, and things can be tracked more quickly and accurately using an

XAI model. EXIF indicates that CAM was made in a child exploitation case at a specific place. Secondary data, such as police records, may be able to tell the XAI who resides in the area. Wi-Fi connection records might indicate that another suspect used the same addresses, and browsing data or conversations could indicate that meetings took place. The XAI will then explain all of these information correlations. Refining might go beyond essential relationships to examine the environment where specific data (such as movies or photos) are stored on devices. AI has previously been used to detect "deep fakes" in pictures and videos using probative analysis.

4. **Receive automated report**: The outcome of the XAI model should be an integrated report with reflections on the case's relevance to the researched items. By referring to specific data points as evidence, this study should provide degrees of confidence in its analysis [7]. Insights into significant persons, locations, times/dates, things, and any correlations it considers relevant based on training data should be included in the analysis. For investigators, visualizing the output will aid understanding and explainability.

9.3 Conclusion

This chapter has demonstrated how various AI approaches are now being applied in various aspects of digital forensics. It has also highlighted typical obstacles, such as the lack of datasets in some domains, the difficulty of presenting findings when particular approaches are employed, and even the difficulty of publishing models with possibly restricted training data inferred from the models. Despite these obstacles, there is much promise for future development. As previously said, this is both in terms of increasing the performance of specific present procedures and the fact that some approaches have yet to be tried in specific regions.

References

[1] David Freire-Obregon, Fabio Narducci, Silvio Barra, and Modesto Castrillon-Santana. 2018. Deep learning for source camera identification on mobile devices. Pattern Recognition Letters (2018).

[2] Luciano, L.; Baggili, I.; Topor, M.; Casey, P.; and Breitinger, F. 2018. Digital forensics in the next five years. In Proceedings of the 13th International Conference on Availability, Reliability and Security, 46. ACM.

[3] Eoghan Casey, Sean Barnum, Ryan Griffith, Jonathan Snyder, Harm van Beek, and Alex Nelson. 2018. The evolution of expressing and exchanging cyber-investigation information in a standardized form. In Handling and Exchanging Electronic Evidence Across Europe. Springer, 43–58.
[4] Simson L Garfinkel. 2007. Carving contiguous and fragmented files with fast object validation. digital investigation 4 (2007), 2–12.
[5] Cinthya Grajeda, Frank Breitinger, and Ibrahim Baggili. 2017. Availability of datasets for digital forensics–and what is missing. Digital Investigation 22(2017), S94–S105.
[6] Weber, Rosina O., et al. "Investigating textual case-based XAI." International Conference on Case-Based Reasoning. Springer, Cham, 2018.
[7] Hoffman, Robert R., et al. "Metrics for explainable AI: Challenges and prospects." arXiv preprint arXiv:1812.04608 (2018).
[8] Stuart W. Hall, Amin Sakzad, Kim-Kwang Raymond Choo. "Explainable artificial intelligence for digital forensics", WIREs Forensic Science, 2021.
[9] Conlan, K.; Baggili, I.; and Breitinger, F. 2016. Anti-forensics: Furthering digital forensic science through a new extended, granular taxonomy. Digital investigation 18:S66–S75.
[10] Qian Chen, Qing Liao, Zoe L Jiang, Junbin Fang, Siuming Yiu, Guikai Xi, RongLi, Zhengzhong Yi, Xuan Wang, Lucas CK Hui, et al. 2018. File fragment classification using grayscale image conversion and deep learning in digital forensics.
[11] Doowon Jeong. "Artificial Intelligence Security Threat, Crime, and Forensics: Taxonomy and Open Issues", IEEE Access, 2020.
[12] David Lillis, Brett Becker, Tadhg O'Sullivan, and Mark Scanlon. 2016. Current Challenges and Future Research Areas for Digital Forensic Investigation. In The 11th ADFSL Conference on Digital Forensics, Security and Law (CDFSL 2016). ADFSL, Daytona Beach, FL, USA, 9–20.
[13] Bojan Kolosnjaji, Apostolis Zarras, George Webster, and Claudia Eckert. 2016. Deep learning for classification of malware system call sequences. In Australasian Joint Conference on Artificial Intelligence. Springer, 137–149.
[14] Quan Le, Oisín Boydell, Brian Mac Namee, and Mark Scanlon. 2018. Deep learning at the shallow end: Malware classification for non-domain experts. Digital Investigation 26 (2018), S118–S126.
[15] Shruti Kapil, Meenu Chawla .2016. Performance evaluation of K-means clustering algorithm with various distance metrics. DOI:10.1109/ICPEICES.2016.7853264.

[16] Muhammad Naeem Ahmed Khan. 2012. Performance analysis of Bayesian networks and neural networks in classification of file system activities. Computers & Security 31, 4 (2012), 391–401.

[17] Mohiuddin Ahmed, Abdun. Naser, and Jiankun Hu. 2016. A Survey of Network Anomaly Detection Techniques. J. Netw. Comput. Appl. 60, C (Jan. 2016), 19–31. https://doi.org/10.1016/j.jnca.2015.11.016.

[18] Ryad Benadjila, Emmanuel Prouff, Rémi Strullu, Eleonora Cagli, and Cécile Dumas. 2018. Study of deep learning techniques for side-channel analysis and introduction to ASCAD database. ANSSI, France & CEA, LETI, MINATEC Campus, France.

[19] Utkarsh MahadeoKhaire, R.Dhanalakshmi 2019 Stability of feature selection algorithm: A review https://doi.org/10.1016/j.jksuci.2019.06.012.

[20] Simson Garfinkel, Paul Farrell, Vassil Roussev, and George Dinolt. 2009. Bringing science to digital forensics with standardized forensic corpora. digital investigation 6 (2009), S2–S11.

[21] Belhassen Bayar and Matthew C Stamm. 2016. A deep learning approach to universal image manipulation detection using a new convolutional layer. In Proceedings of the 4th ACM Workshop on Information Hiding and Multimedia Security. ACM, 5–10.

[22] Ross Brown, Binh Pham, and Olivier Vel. 2005. Design of a Digital Forensics Image Mining System. Lecture Notes in Computer Science. https://doi.org/10.1007/11553939_57.

[23] Luciano, L.; Baggili, I.; Topor, M.; Casey, P.; and Breitinger, F. 2018. Digital forensics in the next five years. In Proceedings of the 13th International Conference on Availability, Reliability and Security, 46. ACM.

[24] Baggili, Ibrahim and Behzadan, Vahid, "Founding The Domain of AI Forensics" (2019). Electrical & Computer Engineering and Computer Science Faculty Publications. 103. https://digitalcommons.newhaven.edu/electricalcomputerengineering-facpubs/103.

[25] Laura Sanchez, Cinthya Grajeda,Ibrahim Baggili, 2019, A Practitioner Survey Exploring the Value of Forensic Tools, AI, Filtering, & Safer Presentation for Investigating Child Sexual Abuse Material (CSAM).

[26] Alison D MacEachern, Divya Jindal-Snape, and Sharon Jackson. 2011. Child abuse investigation: police officers and secondary traumatic stress. International journal of occupational safety and ergonomics 17, 4 (2011), 329–339.

[27] Xiaoyu Du, Chris Hargreaves, John Sheppard, Felix Anda, Asanka Sayakkara, Nhien-An LeKhac, Mark Scanlon. "SoK", Proceedings

of the 15th International Conference on Availability, Reliability and Security, 2020.

[28] Felix Anda, Nhien-An Le-Khac, and Mark Scanlon. 2020. DeepUAge: Improving Underage Age Estimation Accuracy to Aid CSEM Investigation.

[29] Mohamed Faisal Elrawy, Ali Ismail Awad, and Hesham F. A. Hamed. 2018. Intrusion detection systems for IoT-based smart environments: A survey. Journal of Cloud Computing 7, 1 (04 Dec 2018), 21.

[30] Amar Amouri, Vishwa Alaparthy, and Salvatore Dominic Morgera. 2018. Cross layer-based intrusion detection based on network behavior for IoT, In 2018.

10

Blockchain Based Digital Forensics: A Fundamental Perspective

R. Biswas[1*], and S. Biswas[2]

[1]Applied Optics and Photonics Lab, Department of Physics, Tezpur University, Tezpur-784028, India
[2]Department of English, Amguri College, Amguri-785680, India
Email rajib@tezu.ernet.in

Abstract

Blockchain technologies have come a long way as far as smart healthcare is concerned, and has been slowly entering other realms of modern technologies. Of late, smart forensics, or more appropriately, digital forensics, has come under the influence of blockchain technology. Being endowed with decentralization properties, it has become an excellent match as far as integrity and attributes towards the collection of evidence of digital forensics is concerned. As such, a paradigm shift in IoT has taken place with the emergence of the IoT forensic chain. Accordingly, this chapter discusses a fundamental overview of blockchain-based digital forensics. This will be followed by a discussion of special features such as digital forensics incident response, a chain of custody in the context of the IoT forensic chain. We further aim to include practical implementations, with future recommendations.

10.1 Introduction

The Internet of Things has shown its presence in multiple aspects of human civilisation. Being capable of handling a large amount of data and their subsequent processing in the most effective way, IoT has become an inevitable part of modern-day computing, spanning several sectors such as the telecommunication industry, automobile industry, and other allied industries. It is widely established that IoT has been able to curb operational and maintenance costs.

However, other laterals hinder its use to the fullest extent [1–5]. As we know, IoT is characterized by a multitude of sensor nodes where each node is connected through various devices. Consequently, there arises a colossal amount of data for computation/processing, leading to complexity in cyberspace. In the case of an untoward incident such as denial of service (DOS), we can rely on another technological development in data access, namely blockchain technology (BCT). While managing digital witness (DW), BCT can become immensely beneficial, leading to the evolution of state-of-the-art digital forensics and incident response (DFIR) [6–15]. In synchronization with IoT, BCT can facilitate an efficient investigation of cybercrime. With the advent of logs generated by IoT devices, it has become easier to reconstruct the crime scene. However, the integrity of data and admissibility of the same can be effectively executed if we have a chain-of-custody (CC) that looks after and maintains the chronology of events happening during the trial. With the help of an online CC, fraud, abuse, and data falsification can be prevented [12–20]. A schematic of blockchain-based digital forensics is provided in Figure 10.1.

Considering all these, this chapter overviews the inherent features of BCT, which aids in building an efficient DFIR. Starting from the features of IoT forensics basics of DFIR, the implementation of BCT in DFIR is extensively discussed. Additionally, chain-of-custody is outlined, which helps in the chronological documentation of digital witness/evidence during the trial, with further insight on sustainable use of BCT to make it more robust. We also briefly highlight the practical adaptation of these systems and future recommendations.

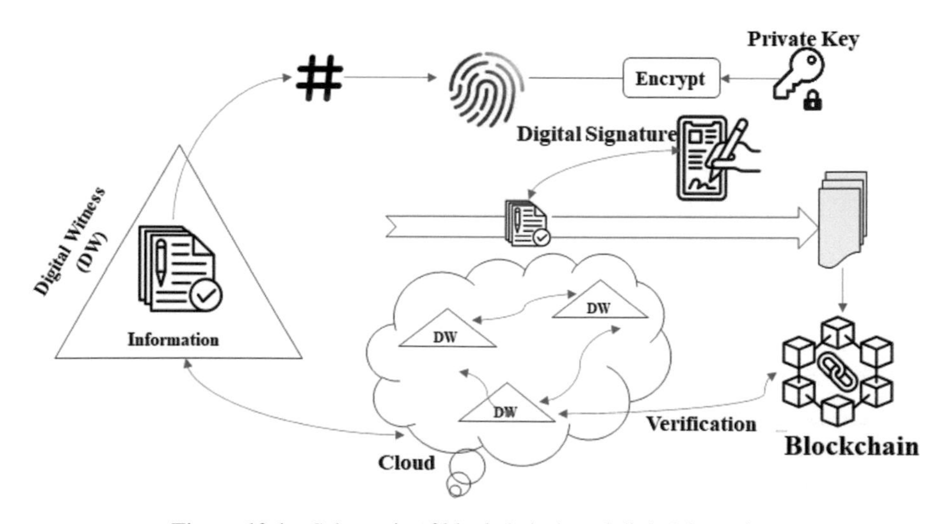

Figure 10.1 Schematic of blockchain-based digital forensics.

10.2 IoT Forensics

The use of IoT technology has risen exponentially over the years. IoT smart devices have now entered several areas of deployment spanning transportation, healthcare, automobile, etc. Consequently, this surge in IoT technology has led to vulnerabilities in many aspects, resulting in cybercrimes [5–15]. The threats and attacks in IoT are rising alarmingly. The primary threats jeopardizing IoT include mischievous users, informal manufacture, external adversary, bad programming, etc. Likewise, IoT attacks include spoofing, privacy breach, overflow for butter, node tampering, denial of service, distributed denial of service, etc. This necessitates a novel investigation approach in dealing with IoT nodes causing the crime. There comes the role of incident response, digital forensics, or incident forensics. There are a couple of approaches to design digital forensics in the context of IoT. Here, we can mention pre-investigative preparedness and the subsequent real-time approach of deploying IoT forensics as the dominant approaches. The former refers to requisite components as a measure of readiness before the occurrence of an incident which calls for preparation, collection, and evaluation. That way, readiness corresponding to management and technology can be mentioned. In a similar manner, the latter covers an automated or conscious investigation related to IoT devices to efficiently manage the diversity of sensor nodes/devices with simultaneous handling of IoT constraints. The following section deals with the fundamentals of digital forensics and incident response (DFIR) with BCT therein, leading to the evolution of secure systems [18-25].

10.3 Incident Response

As the name suggests, incident response considers all the complementary set of occurrences of processes upon identifying an incident. In other words, the activities that ensue in response to an incident can be categorically termed as incident response. It is characterized by easy accessibility with concise communication regarding an effective alert, followed by subsequent resolution of issues. There are specific measures to be reckoned with while developing an incident response methodology in tune with the clients' goals. Supposing the client to be a corporate security professional, the set of priorities will be in synchrony with law enforcement officials. The incident response is likely to cover up the following goals considering the client in mind [4-17].

- Prevention of an incoherent response
- Confirmation or dismissal of occurrence of the incidence

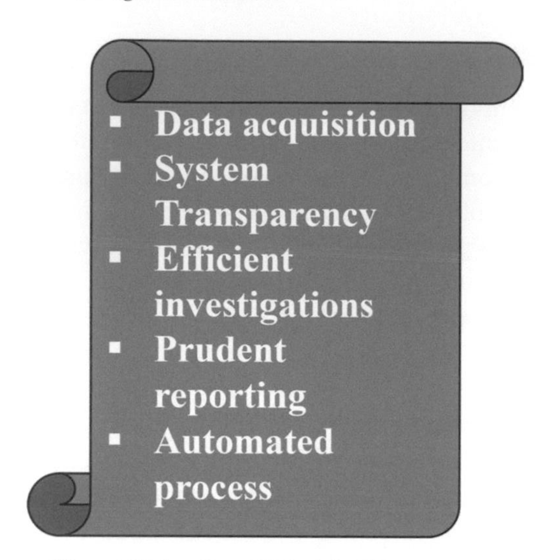

Figure 10.2 Capabilities of a DFIR system.

- Accruing precise information
- Manoeuvring of appropriate retrieval and supervision of evidence
- Strict adherence to law and policy
- Ensuring minimal disturbance to business and network operations.
- Furnishing with accurate information and beneficial recommendations
- Fast assessment and containment thereof.

Now, let us define the important term, i.e., digital forensics. It belongs to that division of computer forensics that caters to appropriate scrutinization of digital components related to either an individual or enterprise by inspecting plausible inappropriate/unlawful actions perpetrated by either the owner or via assorted malicious cyberattacks. Digital forensic technology solutions help clients to support DFIR operations. Figure 10.2 shows some of the capabilities we can expect from a DFIR software solution.

As shown, the DFIR system needs to be equipped with these capabilities. Accrual of multifaced data is imperative. When we say multifaced, we mean data generated by/within many components such as various sensor nodes, devices, or various systems. Likewise, maintaining system transparency is another value addition with which DFIR should be laced. This ensures that actions and administrative functions are visible from every angle. This may lead to other capabilities of the DFIR system, namely, efficient investigations.

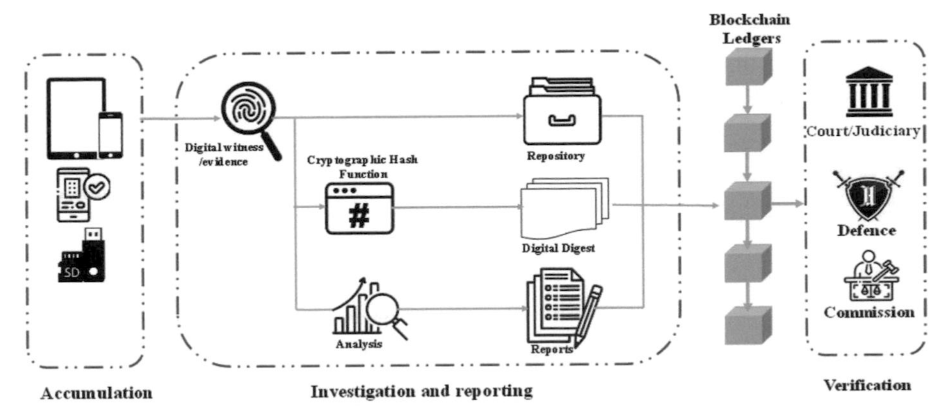

Figure 10.3 Schematic of blockchain-based evidence management as well as incident response.

The efficiency in such cases denotes completeness as well as acquiescence. Once done with investigations, DFIR must be capable of reporting with prudence. This may be effected with powerful articulation/visualization. In addition, DFIR should also engage iterative processes, which should be automatic. The goal is to support allied personnel to explore all instances of artifacts without any intuition rapidly. When smoothly operated, DFIR proves beneficial for small and big enterprises. While evaluating and reducing risk, one must comprehend the phases, as detailed below, which DFIR comprises of [25-36]. Figure 10.3 illustrates the elements related to DFIR and Blockchain.

1. Preparation: Enterprises ready themselves to manage incident response via efficient policies in order and platform software backed by responsive incident managers.
2. Identification: In this second phase, proper diagnosis of the incident occurs, such as the type of attack and the risks involved.
3. Containment: This is another important phase of incident response. Post detection of an attack, the subsequent task for incident managers is containment so that spreading into other vital components can be avoided.
4. Redressal: Once the issue is detected and contained, the next phase is to find the best optimal resolution of the issue via the adoption of forensic analysis.
5. Recovery: After the resumption of services, the enterprise sticks to this phase of IR whereby the continuous monitoring and appraisal thereof are ensured.

6. Broadcasting: Through this phase, the assigned incident manager updates the stakeholders, end-users and public regarding the incident. This, in other words, assures the transparency of IR with the external world.

10.4 Chain of Custody

Chain of custody (CC) turns out to be a very vital parameter for security measures. It refers to systematic maintenance of credentials that spans custody, control, transfer, and an inspection of electronic evidence. CC handles risky phases corresponding to investigation and consequent submission of evidence in court. There must be proper care by the retainer of the digital evidence as tempering by any means renders the latter un-useful during the trial in court [3]. In addition, CC involves assemblage steps such as preservation, packaging, transportation, storage, and inventory [4]. Figure 10.4 exemplifies the block diagram associated with the chain of custody.

In a CC, tampering with digital evidence should be avoided at any cost once it is gathered. A tamper-proof safe location should be ensured for the collected evidence so that fears regarding data tampering can be eliminated, thereby facilitating the production of the same in court without any fear of losing credentials. CC proves very beneficial in identifying the evidence's location, origin, and creator. Apart from this, it can provide the features of the used device. All these pave the way for efficient investigation via evidence leading to clues related to crime and subsequent response adaptation. As stated, at the beginning of CC, keeping digital evidence tamper-proof is daunting. Here, blockchain comes as a saviour. The collected evidence is uploaded into BC to be stored as distributed ledgers that keep their authenticity intact. Each block is encrypted with a hashtag or value. Any attempt of alteration can directly come under surveillance, making the whole process very secure. There is a complete rendering of CC alongside proof during trial [15-25].

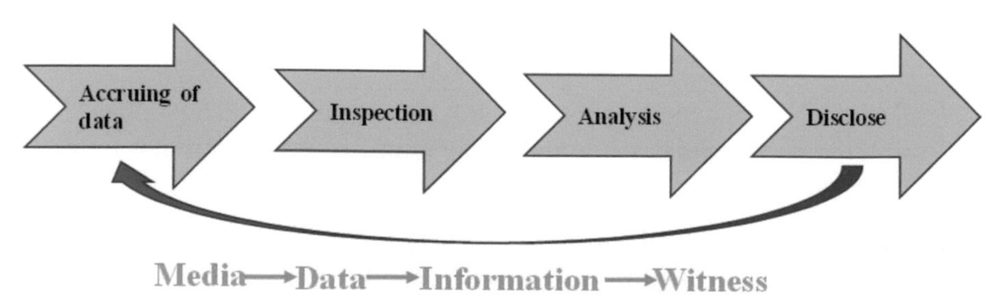

Figure 10.4 Block diagram of chain of custody.

Figure 10.5 A flowchart of BC based chain-of-custody.

Figure 10.5 exemplifies the successive steps to formulate an effective Chain-of-custody. The primary phase goes with the compilation of digital evidence. This may span from DNA analyser outputs, audio-visuals, texts, or images to logs created by IoT devices. All these help in forming a timeline. It is then followed by uploading these digital witnesses into the web/cloud server, ensuring the case files' storage. Correspondingly, the URLs are created. Extraction of these URLs is then executed to form a hash for onward boarding to BC. Along with the hash, the timestamp is also embedded with the extracted URLs, which is finally stored in the corresponding blocks. The timestamp is a marker of authenticity as it helps obtain the actual time of uploading the digital evidence. If there is tampering, this results in chain breakage, thereby alerting a security breach. In the final phase, proof-of-work is mandated, which ensures the integrity of the BC via rehashing the already existing blocks to cross-reference with present information. All these imply that CC can be rationally implemented with the help of BC [20-35]. The user-friendly approach makes it easier to access while outlining its transparency. Both integrity and authenticity can be sustained via issuing unique identification for the user. The eventual mining through which blocks pass makes the whole process of CC secure and tamperproof.

10.4.1 Challenges

CC faces challenges like diminished flexibility and documentation with the rising enormity of data. During the generation of CC, there is a fair possibility

Table 10.1 Blockchain in digital forensics.

Merits	Demerits/challenges
• Tamper-proof feature leads to easy traceability of evidence • Being a decentralized system, blockchain is equipped with better security • Because of intrinsic verification by miners, there exists higher authenticity • Blockchain-based forensics renders better confidentiality than the conventional system as accessibility is provided to only the authorized one • Thanks to the encryption in the form of a hash function, which is not modifiable, thereby ensuring overall integrity	• Legal, regulatory and procedural challenges may arise because of cross-boundary and the diversity of multitude stakeholders abiding by their local rules and governances • Social challenges such as falling prey to losing personal information because of the transfer of data to digital devices exist • Technical challenges spanning the enormity of data generated by several nodes, namely, big data, are found to exist

that evidence may slip from one party to other. Furthermore, it is required that CC should handle them during the trial. To generalize, CC should be capable of furnishing all information related to the case. Simultaneously, it is also imperative for CC to provide source location assisted by information concerning evidence collection procedure. This will facilitate better grasping of the case by the jury members and other law enforcement officers. Table 10.1 lists some advantages and demerits of blockchain in digital forensics.

As evident in Table 10.1, BC possesses some outstanding merits over conventional systems. Rendering better traceability, authenticity, integrity, and security, the use of BC in digital forensics leads to more synergy among the stakeholders. However, at the same time, the use of BC in digital forensics comes with certain demerits. More technically, the demerits are appropriated by challenges that span technical, local, and social challenges, which have been enlisted in Table 10.1.

10.5 Practical Considerations

Mahmud and Rahman [6] showed a practical adaptation of a BC framework-based smartphone app capable of handling a heartbeat in an audio clip. A generated python script aids in absolute point data integrity via BigchainBC. They experimentally demonstrated the application using a virtual private network engaging patients and doctors in conformity with the General Data Protection Regulation (GDPR). This act of manoeuvring can result in a long-lasting impact in BC-based digital forensics, thereby augmenting the security, reduction of cost, etc. Likewise, Lone et al. 2019 [5] conceptualized a forensic-chain based on BC-based digital forensics CC. They showed prototype implementation based on Hyperledger Composer and tested its efficiency. It was observed that the intended implementation yielded robust throughput and optimal use of resources. As per their report, it could be further optimized for end-to-end encryption. Likewise, Lone and Mir, 2017 [7] came up with an idea of a BC framework which could pave the way for forensic adaptations enabling prudent integration and tamper-proofing. As per their claim, the proposed framework could be advantageous in aspects such as traceability of events, fraud prevention with transparent audit trail, acquisition, storage, validation, etc. With gradual development and progress, BC-based digital forensics has slowly been infiltrating different spheres of implementations. Moreover, BC-based digital forensics use advanced software/tools such as Encase 7.5 and Autopsy Media Viewer. In addition, there is an instance of the use of (WIP) blockhub in order to assess untrusted environments.

As evident from the discussions, thc use of BC in digital forensics has been evolving gradually to scale new heights. However, with the advent of more technological advancements, there is every possibility that BC can revolutionize the other prominent emerging sectors with its unique properties. Keeping aside the challenges, one prominent future direction is the synergistic amalgamation of BC with edge devices and cloud computing. Through synchronized assimilation of bigdata analytics with BC, the challenges faced by BC in handling the enormity of data can be dealt with effectively. Thanks to anonymity, non-tampering, and traceability, the blockchain can beef up explainable AI-based systems to counteract more data breaches and better solve complex problems.

10.6 Concluding Remarks

In summary, the chapter overviews the different aspects related to digital forensics. Beginning with IoT forensics, digital forensics and incident response fundamentals are highlighted. We have seen that DFIR can be very effective if the BC is integrated synergistically within it. However, ensuring security of endpoints is a real challenge. These endpoints imply the termination of one node, followed by the initiation of another node. There is a real challenge in ensuring a completely secure system with these endpoints, which increase with the growing volume of data. We see that the deployment of BC leads to more effective handling of the investigation and subsequent trial procedures. Not only that, but the chronology of events also leads to all these forensics being better accentuated by the chain of custody. In addition, IoT smart devices functioning as digital witnesses furnish vital information related to an incident, albeit the IoT device is constrained by its sensing abilities. This requires rigorous adoption of CC so that the admissibility of data generated by IoT devices can be guaranteed. However, more research is required towards the standardization of these IoT devices before being assigned the role of digital witness. It is quite evident that BC smooths the whole process of CC, including security, integrity, and transparency. Additionally, diminution of conflicts and trust build-up remains another feat of BC technology.

10.7 Acknowledgements

The authors gratefully acknowledge the editor for giving the opportunity of writing the chapter. The author also acknowledged the DST-FIST support given to the Department of Physics, Tezpur University.

References

[1] S. Nelson, S. Karuppusamy, K. Ponvasanth, R. Ezhumalai. Blockchain based Digital Forensics Investigation Framework in the Internet of Things and Social Systems, International Journal of Engineering Research & Technology (IJERT) ISSN: 2278-0181, 8 (12), 2020.

[2] H. M. Al-Khateeb, G. Epiphaniou, H. Daly, "Blockchain for Modern Digital Forensics: The Chain-of-Custody as a Distributed Ledger", in Blockchain and Clinical Trial. Securing Patient Data. Advanced Sciences and Technologies for Security Applications, H. Jahankhani, S. Kendzierskyj, A. Jamal, G. Epiphaniou, H. Al-Khateeb, Ed. Cham: Springer International Publishing, 2019, pp. 149–168, ISBN: 978-3-030- 11289-9. DOI: 10.1007/978-3-030-11289-9_7.

[3] S. H. Gopalan, S. A. Suba, C. Ashmithashree, A. Gayathri, V. Jebin. Andrews International Journal of Recent Technology and Engineering (IJRTE) ISSN: 2277-3878, Volume-8, Issue-2S11, September 2019.

[4] R. Yasaweerasinghelage, M. Staples, and I. Weber. Predicting Latency of Blockchain-Based Systems Using Architectural Modelling and Simulation, 2017 IEEE International Conference on Software Architecture, https://doi.org/10.1109/ICSA.2017.22.

[5] A. H. Lone, R. N. Mir. Forensic-chain: Blockchain based dig with PoC in Hyperledger Composer, Digital Investigation 28 (2019) 44e55.
[6] H. Mahmud, T. Rahman. An application of blockchain to securely acquire, diagnose and share clinical data through smartphone, Peer-to-Peer Networking and Applications, https://doi.org/10.1007/s12083-021-01210-6, 2021.
[7] A.H. Lone, R. N. Mir. Forensic-chain: Ethereum Blockchain based digital forensics Chain of custody. Scientific and Practical Cyber Security Journal (SPCSJ) 1(2):21-27, 2017.
[8] H.A. Khateeb, G. Epiphaniou, and H. Daly. Blockchain for Modern Digital Forensics: The Chain-of-Custody as a Distributed Ledger, H. Jahankhani et al. (eds.), Blockchain and Clinical Trial, Advanced Sciences and Technologies for Security Applications, https://doi.org/10.1007/978-3-030-11289-9_7.
[9] M. Pollitt, "A History of Digital Forensics," in Advances in Digital Forensics VI, Berlin, Heidelberg, 2010, pp. 3–15. doi: 10.1007/978-3-642-15506-2_1.
[10] S. Li, T. Qin, and G. Min. Blockchain-Based Digital Forensics Investigation Framework in the Internet of Things and Social Systems, IEEE Transactions on Computational Social Systems, 10.1109/TCSS.2019.2927431.
[11] P. Turner. Applying a forensic approach to incident response, network investigation and system administration using Digital Evidence Bags. Digit. Invest. 2017, 4 (1), 30e35.
[12] K. Wüst, A. Gervais. Do you need a Blockchain? IACR Cryptol. ePrint Arxiv. 2017, 375.
[13] M.S.M.B. Shah, S. Saleem, R. Zulqarnain. Protecting digital evidence integrity and preserving chain of custody. J. Digit. Forensics, 2017, Secur. Law 12 (2), 12.
[14] J. Richter, N. Kuntze, C. Rudolph. Securing digital evidence. In: Endicott Popovsky, B., Lee, W. (Eds.), Proceedings of the 5th International Workshop on Systematic Approaches to Digital Forensic Engineering (SADFE 2010), vol. 2010. IEEE, Institute of Electrical and Electronics Engineers, United States, ISBN 9780769540528, pp. 119e130. https://doi.org/10.1109/SADFE.2010.31.
[15] Y. Prayudi, A. Ashari, T. K. Priyambodo, Digital evidence cabinets: a proposed framework for handling digital chain of custody, Int. J. Comput. Appl. 107 (9).
[16] G. Giova. Improving chain of custody in forensic investigation of electronic digital systems. 2011a, Int. J. Comput. Sci. Netw. Secur. 11 (1), 1e9.

[17] Garfinkel, S.L., Malan, D.J., Dubec, K.-A., Stevens, C.C., Pham, C., 2006. Disk imaging with the advanced forensic format, library and tools. In: Research Advances in Digital Forensics (Second Annual IFIP WG 11.9 International Conference on Digital Forensics). Springer.
[18] A. Alenezi, H. F. Atlam, R. Alsagri, M. O. Alassafi, G. B. Wills, IoT Forensics: A State-of-the-Art Review, Challenges and Future Directions, 2019, DOI: 10.5220/0007905401060115.
[19] H. F. Atlam, A. Alenezi, M. O. Alassafi, A. A. Alshdadi, G. B. Wills. Security, Cybercrime and Digital Forensics for IoT, In Intelligent Systems Reference Library, 2020, DOI: 10.1007/978-3-030-33596-0_22.
[20] Cao X, Xu H, Ma Y, Xu B, Qi J (2019) Research on a Blockchain-Based Medical Data Management Model. https://doi.org/10.1007/978-3-030-32962-4 4.
[21] J. Cosic, M. Baca. A framework to (Im) prove" chain of custody" in digital investigation process. 2010a, In: Central European Conference on Information and Intelligent Systems. Faculty of Organization and Informatics Varazdin, p. 435.
[22] J. Cosic, M. Baca. (Im) proving chain of custody and digital evidence integrity with time stamp. 2010b, In: The 33rd International Convention MIPRO.
[23] Meng W, Li W, Zhu L (2020) Enhancing Medical Smartphone Networks via Blockchain-Based Trust Management Against Insider Attacks. IEEE Trans Eng Manag 67(4):pp. 1377–1386, https://doi.org/10.1109/tem.2019.2921736.
[24] P. Chitti, J. Murkin, R. Chitchyan. Data Management: Relational vs Blockchain Databases, 2019. https://doi.org/10.1007/978-3-030-20948-3 17.
[25] K. Sultan K, U. Ruhi, R. Lakhani. Conceptualizing blockchains: Characteristics and applications, 2018. arXiv:1806.03693
[26] Huckle, S., Bhattacharya, R., White, M., & Beloff, N. (2016). Internet of things, blockchain and shared economy applications. Procedia computer science, 98, 461–466.
[27] Dorri, A., Kanhere, S. S., Jurdak, R., &Gauravaram, P. (2017, March). Blockchain for IoT security and privacy: The case study of a smart home. In 2017 IEEE international conference on pervasive computing and communications workshops (PerCom workshops) (pp. 618–623). IEEE.
[28] Karame, G., & Capkun, S. (2018). Blockchain security and privacy. IEEE Security & Privacy, 16(4), 11–12.

[29] H. Johng, D. Kim, T. Hill, and L. Chung. Using Blockchain to Enhance the Trustworthiness of Business Processes: A Goal-Oriented Approach. pp. 249–252. doi: 10.1109/SCC.2018.00041.

[30] L. Cocco, A. Pinna, M. Marchesi. Banking on Blockchain: Costs Savings Thanks to the Blockchain Technology. Future Internet, vol. 9, no. 3, pp. 25, 2017.

[31] A. Valjarevic and H. Venter. A harmonized process model for digital forensic investigation readiness. in Advances Digital Forensics. Berlin, Germany: Springer, 2013.

[32] M. Cebe, E. Erdin, K. Akkaya, H. Aksu, and S. Uluagac, "Block4forensic: An integrated lightweight blockchain framework for forensics applications of connected vehicles," 2018, arXiv:1802.00561.

[33] L. Caviglione, S. Wendzel, and W. Mazurczyk, "The future of digital forensics: Challenges and the road ahead," IEEE Security Privacy, vol. 15, no. 6, pp. 12–17, Nov./Dec. 2017.

[34] J. H. Ryu, P. K. Sharma, J. H. Jo, and J. H. Park, "A blockchain-based decentralized efficient investigation framework for IoT digital forensics," J. Supercomput., pp. 1–16, 2019.

[35] A. Nieto, R. Roman, and J. Lopez, "Digital witness: Safeguarding digital evidence by using secure architectures in personal devices," IEEE Network, vol. 30, no. 6, pp. 34–41, Nov./Dec. 2016.

[36] R. Biswas, "A Brief Appraisal of Blockchain-Assisted Secured Healthcare" in Prospects of Blockchain in Digital Healthcare, https://doi.org/10.1201/9781003133179-5.

11

Digital Forensics Identity to Improve Transparency in Block Chain Technology Using Artificial Intelligence

K.S. Niraja[1], Fahmina Taranum[2], and Gurram Venkata Siva Nandan[3]

[1]Assistant Professor, Department of Information Technology, BVRIT Hyderabad College of Engineering for Women, Bachupally, Hyderabad, India.
[2]Professor, Department of Computer Science and Engineering, Muffakham Jah College of Engineering and Technology, Banjara Hills, Hyderabad-500034.
[3]Senoir Risk and Reliability Engineer, Phillips-Medisize A/S, Struer, 7600 Denmark
Email: niraja.ksvce@gmail.com; ftaranum@gmail.com; gvsnandan@gmail.com

Abstract

Due to the pandemic situation almost 90% of people are depending on digital technologies to improve their skills, business, education, etc. The usage of digital identities in work spaces, personal lives and other professional related activities has given rise to identity management and access control to manage and secure our identity information. With the help of blockchain-based identity management solutions users can take control of their own identity. The large amount of data stored in the public network leads to data leakages. Due to lack of confidentiality, slow and speed scalability especially when we store large amounts of data cannot precede in a single block which causes degradation of data. Even though the system performance can be analyzed with accuracy and integrity still there is a need to refine identity management in blockchain technology in digital forensic as it is built on a platform using protocols and hashing functions like private/public keys where the owner's data is hidden. Artificial intelligence (AI) usage has risen greatly in every field, and with the help of AI and blockchain the identity management in digital forensic can be improved to eliminate duplication of work to increase transparency to improve speed of service. In this chapter we discuss possible identity management techniques like protocols and algorithms to enhance the possible solutions in digital forensics.

11.1 Introduction

The collaboration of IoT devices, blockchain technology with digital forensics, helps to provide confidence to all participants, making the identification of a digital forensics investigation more transparent. Forensics refers to assessment and argument on analysis performed on public data, and digital media refers to computers, servers, electronic gadgets or any network. The aim of digital forensics is to use the best techniques and tools for solving crimes using evidences from digital media [1]. The motive is to uncover electronic crime data by carrying out an investigation on the evidence in a court of law.

11.1.1 Digital forensics

The process of the forensics discipline is depicted in Figure 11.1 which concentrates on recognizing, fetching and processing, analyzing and reporting, and storing the data electronically [2]. The aim is to make these operations transparent using advanced techniques of artificial intelligence.

11.1.2 Standards used to practice digital forensics

The guidelines framed to initially handle digital evidence are defined using four phases [3], as depicted in Figure 11.2.

a. Identification: This deals with the documentation and recognition of relevant evidence based on the value and volatile nature of evidence.

b. Collection: The integrated group of all digital devices that potentially contains related evidentiary value of data. This collection is stored in the forensic lab and is named static acquisition.

c. Acquisition: This is an act of learning or developing of a skill to acquire an object, obtained by retaining integrity. Dynamic acquisition works on collecting volatile and non-volatile data from live executions and is capable of interfering with the functionality of the control system. The live data should be converted to an understandable computer format.

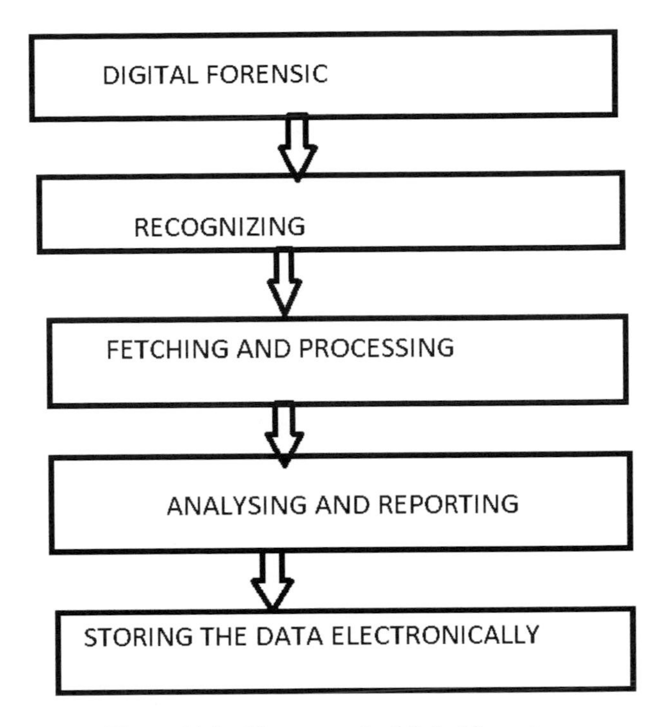

Figure 11.1 Flow control of digital forensics.

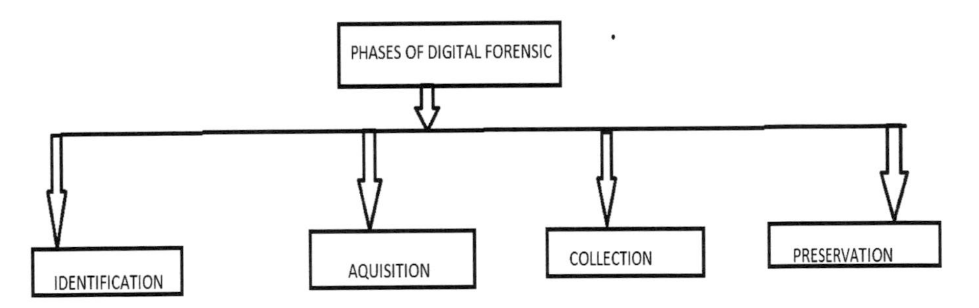

Figure 11.2 Phases of digital forensics.

d. Preservation: The integrity of the devices is preserved by using the norms of human rights and legal constraints.

11.1.3 Summarization of the challenges in existing digital forensics investigation as depicted in figure 11.3

a. Trustworthy: Applying and improving the trustworthiness of an evidence item in digital forensics is difficult.

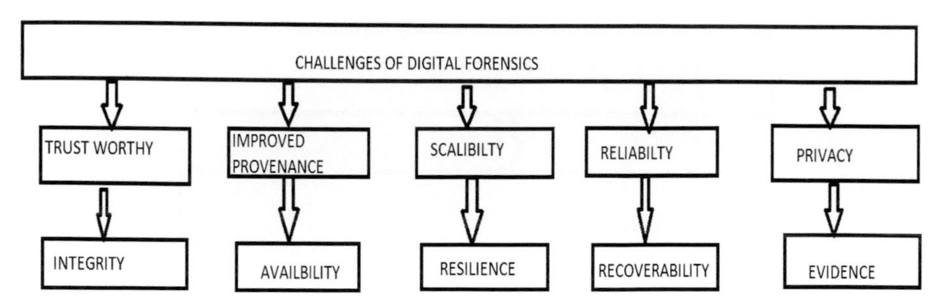

Figure 11.3 Challenges of digital forensics.

b. Integrity: Digital investigation requires continuous integrity checking.

c. Improved provenance: All the evidence is validated using hash functions, which is a difficult and challenging task.

d. Scalability: This is to decide the capacity of data that can be stored at each level. In digital forensics each node can hold 1000 pieces of evidence.

e. Availability and resiliency: Each node must have an accurate and complete copy of the hash tree, to make it resilience.

Other challenges include: defining the framework to cater to new challenges in anew environment and ensuring the reliability, availability, recoverability of dynamic evidence and privacy concerns.

11.1.4 Blockchain

Blockchain technology is a distributed archive system, used to store linked records in a decentralized database template for a peer–peer network. Electrical evidence is a constituent of almost all criminal occurences and digital forensics maintenance is crucial for law enforcement examination [4]. Blockchain makes artificial intelligence more self-sufficient, self-directed and trustworthy, and artificial intelligence can prompt blockchain toward intelligence.

Blockchain has issues concerning energy consumption, security, scalability, privacy, immutability, reliability, decentralization, anonymization, and efficiency. It helps to guarantee the credibility of originality, trace, and track to analyze behavior, and audit credibility for trustworthy and transparent decision making of AI.

11.1.5 Artificial intelligence

Artificial intelligence is used to discover risk and carry out future predictions of crime by training the system with huge data. This helps to solve problems in dealing with interpretability and effectiveness. The critical or criminal patterns of risk are easily identified on huge data using artificial intelligence techniques. In realistic scenarios, artificial intelligence works with three key elements in collaboration: algorithms, power consumption in computing, and data, whereas the blockchain can deal with the hurdle of data representation and realize the easy flow of algorithms in integration with computing power and data resources[5].With the increase in demand of intelligent analysis on large or complex datasets to deal with crimes, artificial intelligence is used to make the process easier and faster. The artificial intelligence supports in optimizing the creation of blockchain by making it more secure, accurate, fast, and energy efficient.

11.1.6 Internet of things(IOT)

The Internet of Things designates physical entities that are implanted with sensors, possess handling ability and are integrated with advanced technologies to communicate over the internet by exchanging data without any involvement of humans [6]. The Internet of Things is stitching the world together with smarter and more responsive devices, and merging digital and physical entities to produce efficient output. IoT devices will collect the data by communicating with the RFID, web camera, and IPV6.

Features: It is used for wider connectivity, real time analytics, uses sensors to sense faults and report status, helps to connect technology, provides good end point management, and is easily integrated with other models. The reason for integrating with AI is to make thing smart, enhance the speed, precision, and effectiveness of human efforts, and to enhance the quality and accuracy of risk detection.

Objectives of IoT are inclusiveness, continuous integrity, an interference-proof environment, full derivation and traceability. Figure 11.4 depicts the overview of IoT, blockchain, and AI applied to digital forensics.

11.2 Literature Survey

The authors in [7] use digital forensics to examine contracts related to crime for digital devices like laptops, tablets, mobile phones, etc. Many committees work together to frame a template to fit the regulatory strategies of digital

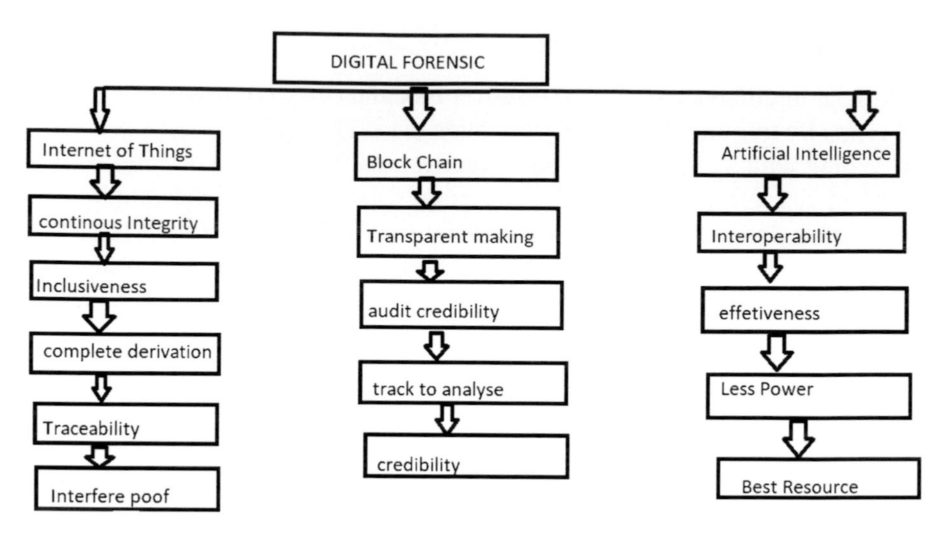

Figure 11.4 Objectives of IoT, Block-chain and AI.

forensic. The Association of Chief Police Officers (ACPO) control the digital investigations by following the principles listed below

Principle 1: The changes in the data made by authentic agents must be reliable and synchronous with the court.

Principle 2: Only competent people are admitted to access the original data upon providing evidence for the need and necessity of their actions.

Principle 3: The involvement of a third party is initiated to do the audit trail for validating the evidence and preserving the same.

Principle 4: The designated in-charges take the sole responsibility of safeguarding and abiding by the law.

The UK Forensic Science Regulator (UK-FSR) presented ISO17025, a countrywide recognized identified practical standard for analyzing and validating the approaches and techniques. This standard defines the quality and is required to be followed in forensic labs.

The authors in [8] use digital forensics in collaboration with processes like data acquisition, preservation of data, analyzing data, and reporting on statistics regarding a crime investigation. The article discriminates the scientific investigations from manual ones using the research conducted on digital forensics. Scientific research is growing continuously. Digital forensics creates pieces of evidence for the court of law to solve criminal cases.

In [9], the enhancement of IoT with added security concepts for streaming videos helps to minimize security breaches. The author has proposed compression techniques to regulate the protection strategies. The aim is to detect forgeries using automatic detection process.

The authors Erika and Hanif et al. in [10] explain tools like SENS Investigative Forensic Toolkit (SIFT) Workstation, which defines a group of open foundation forensic tools disseminated by the SANS Institute. SANS grades tool physiognomies such as file system, image proof, and partition table support. The investigation generated from this tool helps to analyze reports in court.

The Computer Forensic Tool Catalog (CFTC) is an online selection system tool developed by the National Institute of Standards and Technology. This tool works on the logic of hash algorithms to generate evidence.

The proposal is to work by designing an expert system with forward chaining that works on six characteristics and three functionalities. Characteristics like memory requirement, processing speed, output format, required skills, focus and cost are used. The functionality on which it works includes low, moderate, and high.

The authors in [11] explain that the fast expansion of IoT introduces many safety challenges. With the increase in the demand of forensic procedures, the need to explore and invent new approaches to deal with complexities in innovation increases. The work has highlighted the operational interventions to resolve crime related issues quickly. Incorporating digital evidence in an IoT environment is a source of potential challenge in forensics science [12]. The methodology for identifying and classifying connected objects in search of the best evidence to be collected focuses on technical and data criteria to successfully select the relevant IoT devices. This chapter discusses the difficulties of identifying locally connected objects and prioritizing that selection of evidence within the IoT infrastructure.

The authors in [13] explain the benefits of digital forensics. A popular area for research in digital forensic science is seeking a standard methodology to make the process more robust, efficient, accurate, and fast. The steps used are: acquisition, identification, evaluation and admission. With the involvement of cloud computing for companies like Amazon, Google, Microsoft and IBM, the shifting of these services to DF is considered to be profitable as it assists law enforcement personnel to progress their caseload.

A meticulous methodology is a procedure of examination for conducting digital forensic investigations. Researchers have proposed many new methods, approaches, techniques, and tools in the examination process to deal with new challenges encountered in the new world of modern investigations[14].

A data reduction and data mining framework is used to pre-filter the increasing volume of data to reduce storage requirements. With the increasing demand of digital evidence, the amount of work required to propose solutions to deal with issues and challenges for safe prediction increases. This is taken as a reason to generate multiple examinations for critical problems with deep historical case study and intelligent analysis.

The authors in [15] from VTO Laboratory and Salford University have discussed the issues and challenges related to practitioners, manufacturers, and legal authorities in accessing forensics. They have worked together to reduce the gap of the forensics device connectivity and its association with IoT.

The aim of the authors in [16] is to highlight the thorough valuation of the activities and practices used by mobile devices to do a significant survey of over seven years of data on mobile forensics. The work has highlighted the flow in the trend of mobile forensics and its frequent modifications to adapt to the changing ecosystem.

11.3 Role of IOT and Blockchain to Improve Transparency in Forensics

The objcctives of IoT blockchain and AI applied to obtain transparency in digital forensics is depicted in Figure 11.4. The functionalities used in collaboration to improve the identity of digital forensics is listed as follows:

i. The increase in cybercrime has initiated and increased the demand for IoT devices. IoT devices are inventions of 5G communication and a decentralized blockchain technology holds copies at multiple locations; these characteristics are making them a part of contemporary life [17]. Digital forensics that deals with IOT related cybercrime investigations, which are completed with the help of sensors, devices for connection, and data back-up, are termed as the IoT forensic chain. This helps to make data representation transparent as it offers authenticity, traceability, resilience, and trust over distributed entities; simultaneously it helps to track the derivation related to evidence.

ii. The blockchain helps to gain secured, immutable blocks with increased transparency, auditability and accountability of data. The blockchain technology can propose forensic applications with significant reimbursements for the complete procedure of digital forensics exploration including data collection, stabilizing consistency, evidence certifying, data investigation, and demonstration for finding the best presentation pattern[18]. Blockchain helps to improve the transparency in the very initial stages by examining and identifying the accurate data sources used in the investigation to improve efficiency and reduce the cost of investigation.

iii. Ensuring the integrity of specific evidences is a basic constraint in IOT devices, which is taken as a challenge and is solved using cryptographic functions with constant integrity validation of specific evidence items[18]. These evidence items are represented in blockchain to pledge the full derivation. IoT devices generate the timestamp information, which is converted as the node and added to the blockchain network to improve security.

iv. To trace a blockchain, there is some potential privacy issue, which is solved and protected by using encryption [18]. The role of IoT is to discover easy tractability to avoid glitches and restrict access to all confidential recorded information.

v. The decentralized system feature of IoT supports generating complete evidence, and blockchain provides trust to carry out investigations and maintain transparency [18].

11.4 Framework of Digital Forensics

The framework used traditionally to define digital forensic includes:

a. The Digital Forensic Research Workshop (DFRWS) model, which is a non-profitable organization to discuss the framework of DF.

b. The integrated digital investigation process (IDIP), which helps automatic interventions to organize all hidden data to expose key evidence and discover hidden connections for presenting and investigating DF.

c. The enhanced integrated digital investigation process (EIDIP) helps to work on cybercrime related investigation with amplified computer and network misuse.

d. The integrated digital forensic process model (IDFP), which works on the four pillars to generate the investigation using preservation, collection, examination, and analysis phase.

11.5 Proposed System

The aim of the proposed system is to identify the benefits of digital forensics to improvise the application using IoT and blockchain technology and is depicted in Figure 11.5. The process is to collect the input from the multiple input devices. The input is encrypted to enhance the security of the message. The evidence initialization is represented by keeping a backup in the repository. The identity verification is done by collecting supportive evidence from the repository. The digital forensics is compiled on the digital data to validate the processed evidence. The analytics process helps to do an in-depth study and examine the electronic stored data. The statistics are used to represent the graphical output of the analytics. The result generated using this process helps to support the judgmental proceedings related to critical civilian issues and criminal investigations. The timestamp stores the date and time of the evidence generated. Autopsy is a program used to simplify the installation

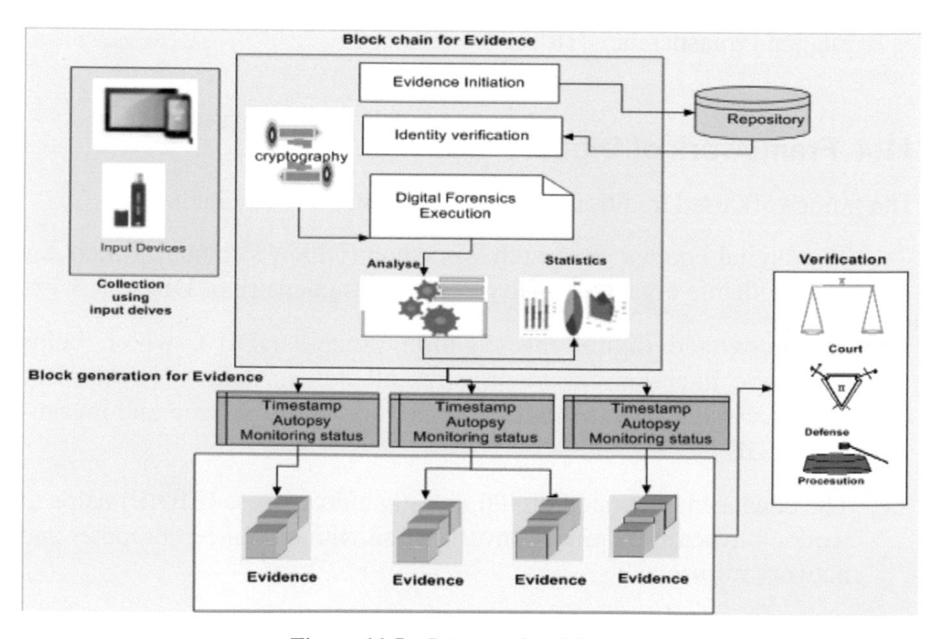

Figure 11.5 Integrated architecture.

and deployment of multiple open source platforms and a plug-in is used to graphically represent the results generated from huge data extracted from the processing of evidence, which helps the examiners to highlight the relevant portions of data.

Digital forensics, on the other hand, serves both law enforcement and enterprise businesses looking to investigate attacks on their digital property. In law enforcement, digital forensics specialists are increasingly becoming in higher demand as cybercrime rises. In corporate security, businesses rely on these professionals to ensure their business and customer data remains secure and useable. The component monitoring status is used to check the status of acquisition, authentication, and analysis for altered, damaged or recovered data. These blocks of evidence are passed for verification to validate and resolve the evidence in the best possible way [18].

With the fast transforming world, the role of these latest techniques gives a perfect solution to solve crimes. According to the National Institute of Standard and Technology, USA the experimentation done towards these enhanced strategies helps to adapt to the advancing trends of innovation and thereby act as a medium to improve the standard and quality of living.

The proposal of integrating the methodology as proposed by the authors in [19] is depicted in Figure 11.6. The aim of the proposal is to work on the phases like investigation, identification, detection, and localization. Hindustan Aeronautics Limited (HAL) has a tool used to upload the research work for direct scientific communication related to academics. Detection operations use complex and standard protocols to proceed with communication, which helps to support spectral scattering of messages using a direct spread spectrum or FDD approach to switch the frequency and modulate the signal for generating encrypted communication. The received signal strength (RSS) is used to receive, recognize, and locate the devices using sensors. The examination could be either to locate related to Wi-Fi indoor localization or outdoor devices [20].

The single receiver method scans the complete crime scene with a multi-antenna sensor with high- speed. For example, in drones the multi-sensor network consists of dimensions from fixed and mobile points. This system consists of sensors with accelerometers, external capabilities and gyroscopes. Under external capacitors the components used include Global Navigation Satellite System (GNSS), wave propagation, and thermal signatures, which help to contextualized the parameters of space and time in terms of increasing performance. This proposal helps to resolve the difficulty encountered in identifying the locations of objects to prioritize the selection of evidence for the IoT infrastructure.

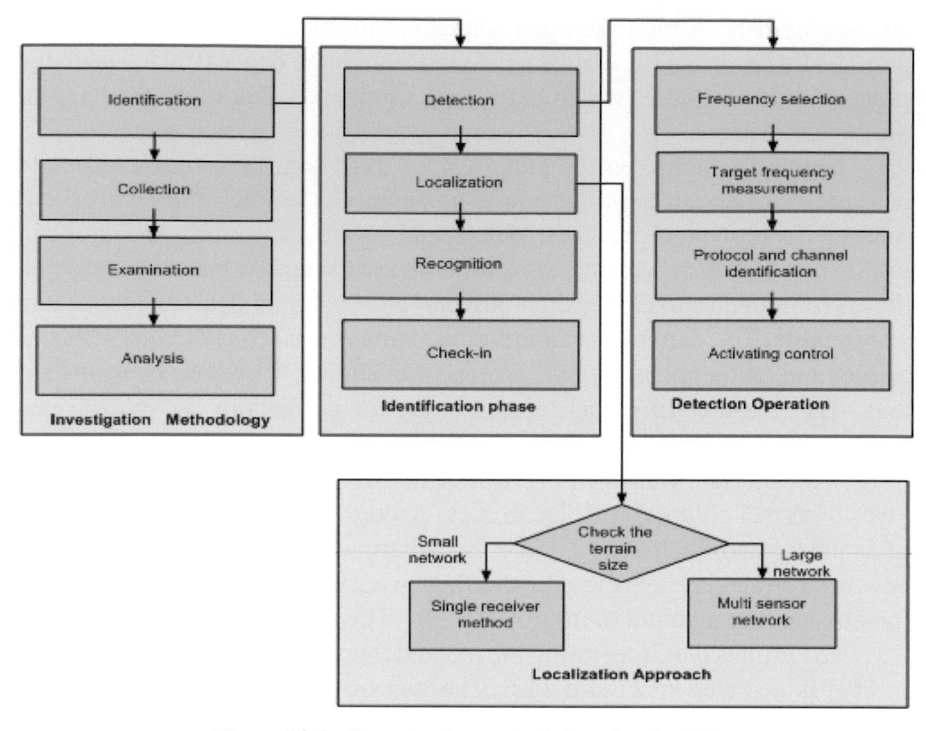

Figure 11.6 Investigation methodology for the IoT.

IoT with forensics facilitates the examiners to acquire intelligence in digital targets pertaining to cyberattacks. This investigation helps to rule out the manual errors and increases the truthfulness of scientific evidence.

11.6 Hardware and Network Challenges in Digital Forensics

The new methods for digital forensic applications involve areas like file management systems, data that is stored in different file systems analysis, network traffic along with log files, which are major challenging tasks which have to be maintained in efficient way [21]. Different network challenges like data have become complex where it supports high volumes of data along with speed. Cloud resources with the use of social networks have given rise to more tools, along with more time needed for reconstruction of digital forensics.

Privacy preservation is also one of the most challenging tasks where data is collected from different social networks, and the data is stored in a different encrypted format to avoid cybercrimes. Digital forensics is applied in many fields: data mining, finance and networking. Blockchaining is also one of the enhancing areas with more secure features that can be effectively implemented along with AI machine learning algorithms for advance prediction and can provide more secure features. Digital forensics applications were initially widely used in industries and business applications but it has spread its usage to even academic areas, and with the introduction of IoT devices its usage is even greater. Botnet's usage has been widely improved nowadays and we need to develop and design more advanced algorithms where we can efficiently improve the usage of digital forensics technology in every aspect.

11.7 Conclusion

The chapter gives a brief overview of digital forensic approaches and its framework. The different existing frameworks along with the proposed framework is discussed. The framework along with blockchain technology gives ample ideas of where it can be used to improve transparency and privacy preserving policies. The usage of different standard protocols, along with possible hardware and network challenges is discussed. Blockchain and AI eliminate the problem in conventional digital forensics techniques. Digital forensics applications are widely used in all fields, and can be improved even further with the help of more secure and standard protocols, frameworks, etc. Botnets have been widely improved to provide more secure and privacy preserving policies and we need to enhance even more algorithms in future.

References

[1] Philip Anderson, Dave Sampson and Seanpaul Gilroy, "Digital investigations: relevance and confidence in disclosure," Sciencegate, Springer September 2021, DOI: 10.1007/s12027-021- 00687-1.

[2] The Digital Process. https://www.open.edu/openlearn/science-maths-technology/digital- forensics/content-section-4.1. Accessed 9 July 2021.

[3] Ioannis Memos Bagkratsas, Nicolas Sklavos, "Digital Forensics, Video Forgery Recognition, for Cybersecurity Systems," Euromicro Symposium on Digital System Design, 2021.

[4] Erika Ramadhani, Elyza G. Wahyuni, Hanif R. Pratama," Design of expert system for tool selection in digital forensics investigation," IOP Conference Series Materials Science and Engineering , vol. 852, 2020, doi:10.1088/1757-899x/852/1/012137.

[5] Francois Bouchaud, Gilles Grimaud, Thomas Vantroys and Pierrick Buret, "Digital Investigation of IoT Devices in the Criminal Scene, "Journal of Universal Computer Science, vol. 25, no. 9 (2019), 1199–1218.

[6] Xiaoyu Du, Nhien-AnLe-Khac, Mark Scanlon, "Evaluation of Digital Forensic Process Models with Respect to Digital Forensics as a Service," (2017). ArXiv preprint, arXiv: 1708.01730.

[7] Steve Watsona and Ali Dehghantanhab, "Digital forensics: the missing piece of the Internet of Things promise," Computer Fraud & Security, Volume 2016, Issue 6, June 2016, Pages 5-8, Elsevier.

[8] Barmpatsalou, K., Damopoulos, D., Kambourakis, G., Katos, V.," A critical review of 7 years of Mobile Device Forensics," Digital Investigation vol. 10, no. 4, pp. 323–349 2013.

[9] R. Montasari, "An Overview of Cloud Forensics Strategy: Capabilities Challenges and Opportunities", *In Strategic Engineering for Cloud Computing and Big Data Analytics*, pp. 189–205, 2017.

[10] Caviglione, S. Wendzel and W. Mazurczyk, "The Future of Digital Forensics: Challenges and the Road Ahead", *IEEE Security & Privacy*, no. 6, pp. 12–17, 2017.

[11] A. Pichan, M. Lazarescu and S.T. Soh, "Cloud forensics: Technical challenges solutions and comparative analysis", *Digital Investigation*, vol. 13, pp. 38–57, 2015.

[12] D. Lillis, B. Becker, T. O'Sullivan and M. Scanlon, "Current challenges and future research areas for digital forensic investigation", 2016.

[13] J. Jang-Jaccard and S. Nepal, "A survey of emerging threats in cybersecurity", *Journal of Computer and System Sciences.*, vol. 80, no. 5, pp. 973–993, 2014.

[14] K. Ruan, J. Carthy, T. Kechadi and I. Baggili, "Cloud forensics definitions and critical criteria for cloud forensic capability: An overview of survey results", *Digital Investigation*, vol. 10, no. 1, pp. 34–43, 2013.

[15] P. Van Kessel, "Is cybersecurity about more than protection?" *Ey Glob. Inf. Secur. Surv. 2018–2019*, 2019.

[16] B. Carrier and E. H. Spafford, "Getting Physical with the Digital Investigation Process", *Int. J. Digit. Evid. Fall*, 2003.

[17] V. Baryamureeba and F. Tushabe, *The Enhanced Digital Investigation Process Model Venansuis Baryamureeba and Florence Tushabe*, 2004.

[18] M. Kohn, M. S. Olivier and J. H. P. Eloff, "Framework for a Digital Forensic Investigation", *Communications*, 2006.

[19] March Hevner, Park and Ram, "Design Science in Information Systems Research", *MIS Q.*, 2004.

[20] Seamus Ó Ciardhuain, *An Extended Model of Cybercrime Investigations*, vol. 3, no. 1, pp. 1–22, 2004.

[21] B. Carrier and E. H. E.H. Spafford, “An event-based digital forensic investigation framework”, *Proc. 4th Digit. Forensic Res. Work.*, pp. 11–13, 2004.

12

Forensic Analysis of Online Social Network Data in Crime Scene Investigation

S. Saranya[1], and G. Usha[2]

[1]Research Scholar, Department of Computing Technologies, School of Computing, SRM Institute of Science and Technology
[2]Associate Professor, Department of Computing Technologies, School of Computing, SRM Institute of Science and Technology
Email: saranya.it78@gmail.com; ushag2@gmail.com;

Abstract

Online social networks (OSNs) have become a part of our everyday lives. The information created in OSNs will be examined when criminals misuse this information. Criminological evidence was assembled by analysing OSNs using online digital forensics tools. Analysing OSNs will help computerised criminology investigators forecast, identify, and prevent crimes in various fields. In mobile device forensics, forensic methods are used to recover digital evidence from mobile devices using forensically sound techniques. This chapter discusses the crimes on social networking sites, phishing, cyberbullying, baiting with a link, and digital evidence analysis. We discuss how the user can securely use OSNs. Moreover, we discuss how forensics departments can create evidence and submit it to the law.

12.1 Introduction

Social media creates a tremendous amount of information. The assembled data is used by electronic devices like mobile phones, tablets and IoT devices. These devices are connected to the public internetwork, and the data associated with devices is easily misused. [1]. Organised crime groups (OCGs)

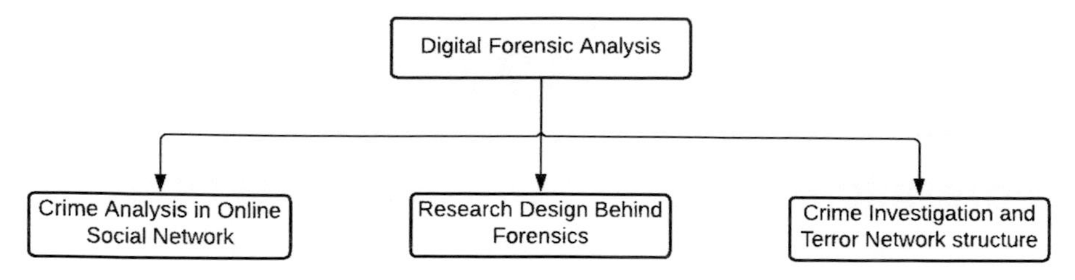

Figure 12.1 Flow diagram of digital forensic analysis.

often work in well-organised international or local groups for cheating and misusing the information in online social network [2].

Computerised criminology is expanding because digital forensics developments are growing fast [3][4]. Existing examinations, like applications in advanced criminology, centre around document frameworks, log records, network traffic, data sets, and cell phones and tablets, are submitted to courts for evidence [5].

The information left on cell phones can contain plenty of data. For example, pinpointing past correspondence is used for stealing digital reports [6][7]. Law enforcement officials request specific data from administrators. However, they are generally not obligated to cooperate with requests from other countries [8][9][10]. Furthermore, the increased usage of encryption makes data security and analysis extremely difficult. At the same time, counter-legal devices are quickly obtained via the Internet. As a result, digital legal sciences are consistently lagging behind the current state of innovation. Figure 12.1 depicts the digital forensic analysis.

Crime analysis of online social networks, research design behind forensics, crime investigation and terror network structures are the three domains to be discussed under digital forensic analysis.

12.2 Crime Analysis of Online Social Networks

This section discusses the current status of online social networks (OSNs) and the effects of these networks on explicit people and societies. It also discusses the abuse of OSNs in carrying out cybercrimes going from financial wrongdoings, digital harassment [11], and spreading of fake news, to the enrolment of radicals [12] [13]. Cybercriminals will generally take considerable time to safeguard their identities, and specialists suggest pursuing cybercrimes committed through a proxy server.

Web-based media clients: The web-based media client is used by end users in the social media. The web-based media client does not give an idea of how

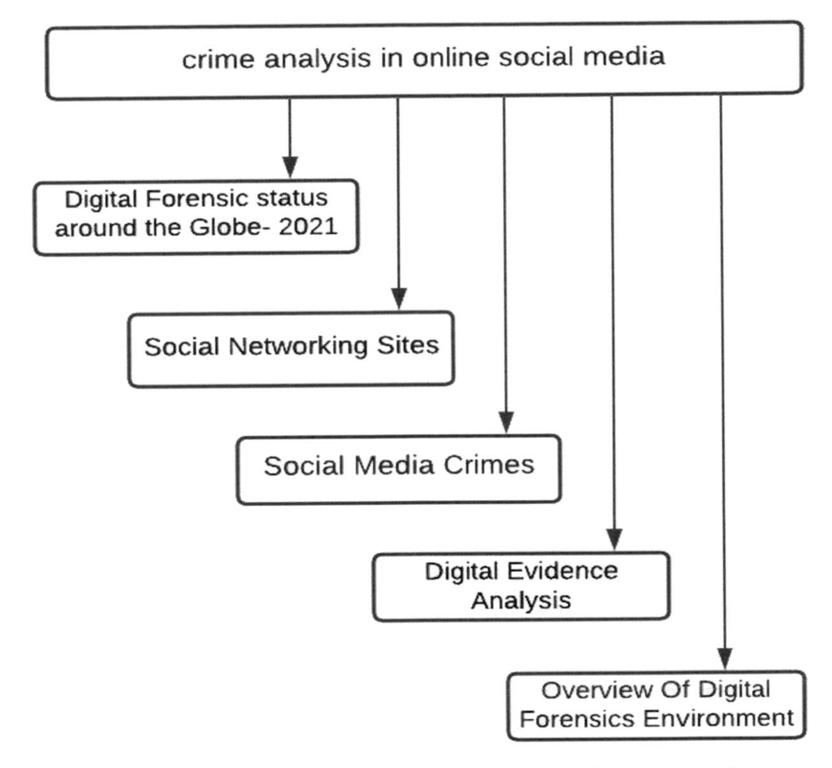

Figure 12.2 Flow diagram of crime analysis in online social media.

cybercriminals misuse social media. [14]. Some web-based media records might address beings, organizations, places, and different "non-human" substances. At the same time, a few groups may likewise oversee more than one web-based media account on a similar stage. Thus, the concepts behind this section change how to talk about web-based media client numbers to clarify that the figures do not address "individuals" [15] [16] [17]. The source used for web-based media clients in China, which has brought about certain progress to verifiable figures, including worldwide aggregates, has to be changed.

Web-based media records: A technology-based domain that allows users to create, post, collaborate and share content. Web-based media allow two-way communication, whereas traditional entertainment like television and radio is one-way communication [15] [16] [17]. Figure 12.2 shows a crime analysis on online social media.

Digital forensics has the option to re-compute web client figures for consecutive years depending on similar sources [21][22][23][24] and this is utilised for the present year, so the yearly and quarterly development figures as remembered for the current year's reports precisely reflect how much web client numbers have changed over the long run. If it's not too complicated, refer to the growth of social media statistics and analyse any changes at any point over time.

12.2.1 Digital forensics status around the globe – 2021

The following figures give the digital usage percentage of unique mobile users, Internet users and active social media users in the total population of people around the globe [25]. Figure 12.3 depicts the digital usage percentage around the globe.

A portion of the vital topics to search for in the current year's reports include:

i. Changes in how individuals look for data and brands.

ii. The advancing socioeconomics of online crowds.

iii. The quickly developing significance of online business.

iv. Why portable is fundamental? By any means is not the only answer.

v. Why do we need to change the measurements that guide our online media "blend"?

Figure 12.4 depict global digital growth in the form of a percentage of unique mobile users, Internet users and active social media users in the total population of people in the globe[26][27][28].

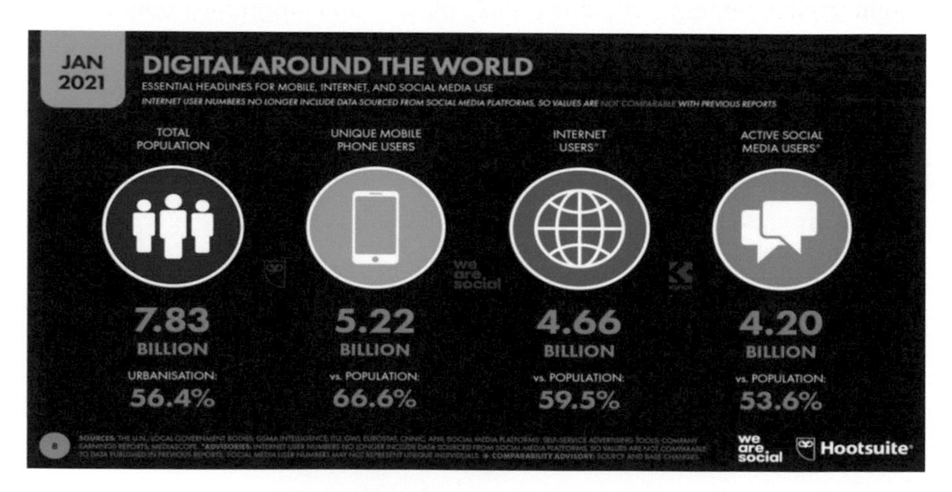

Figure 12.3 Digital usage percentage.

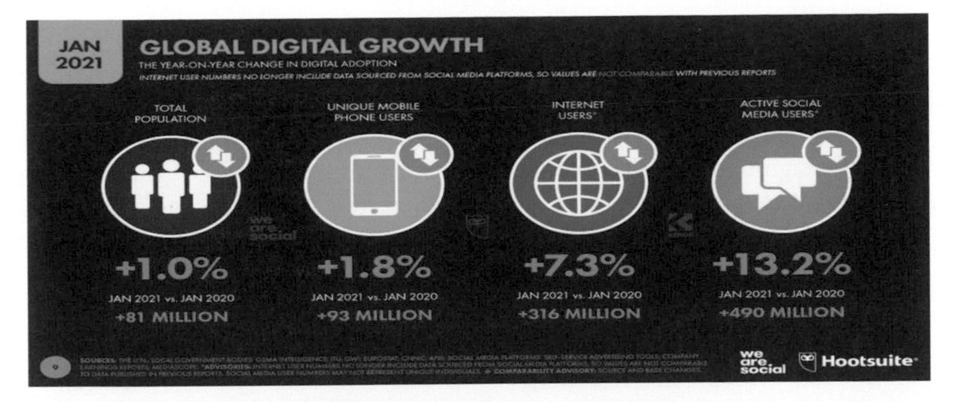

Figure 12.4 Global digital growth.

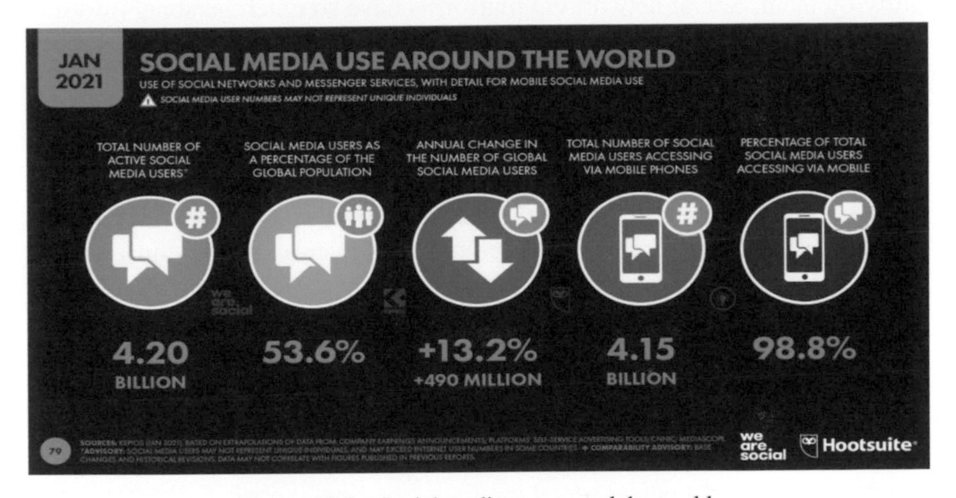

Figure 12.5 Social media use around the world.

Accordingly, web client numbers have likely become more than 7%. These figures will need to be updated once everyday existence (and exploration) resumes[29][30].

Figure 12.5 depicts social media usage around the world. The total number of active social media users is 4.2 billion. Social media users as a percentage of the global population is 53.6%.

Web-based media [31] client numbers expanded by 13% over the previous year, with a large portion of n new clients, taking the worldwide client to practically 4.2 billion by the beginning of 2021.

Considering all aspects, more than 1.3 million new clients joined online media during 2020; 15.5 new clients every second.

Notwithstanding critical changes in computerised practices because of COVID-19, individuals state that they spend a generally similar measure [32] of time daily on online media today as they did last year.

GWI information shows that the day-by-day average has expanded by the more significant part of an hour in recent years. Presently, regular clients use web-based media daily for 2 hours and 25 minutes; about one full day of their lives every week [33] [34].

12.2.2 Social networking sites

Social networking sites are easy to operate and use. There are currently over 3.397 billion social media users, who spend an average of 116 minutes daily on them. Social networking platforms have so much personal information freely available [35] that they have become a hotbed of illicit activity. However, there are investigations to bring justice to the victims and prevent future incidents [36][37]. Investigators teach how to gather social media forensics evidence and forensic examinations of mobile social networking apps [38].

In the last few years, online communication in social networking has seen a massive transformation. Three hundred twenty million more people used social media during September and October 2018 than in September 2017. There are new social network members every 10 seconds. There are 60 billion messages sent every day via WhatsApp and Facebook Messenger.

12.2.2.1 Statistics on social media

People live in an era where social media is growing exponentially. The number of new Facebook users each day is 500,000. This equates to six profiles being created per second [40] [41]. When we are talking about social media, why are we not talking about YouTube? The daily mobile video views on YouTube are approximately 1148 billion. 500 million tweets are sent each day and 326 million month-to-month live customers on Twitter! 74.7 million blog posts are written by WordPress users each month, which is excellent content.

500 million tweets sent each day

12.2.2.2 Various types of social networking sites

Social media is a term that we have all heard about. Many people are unaware that there are other social media networks other than Facebook, Instagram, Twitter, Snapchat, and WhatsApp. The most important goal of social media platforms is to categorise them. A list of different types of social networking sites is provided below.

Sites on social media

Networks such as social networks, also known as "connection networks", promote the sharing of information and ideas between individuals and groups over the internet [42] [43].

Uses: To virtually associate with other persons or brands.

Examples: Facebook, Twitter, WhatsApp, and LinkedIn are just a few examples.

Networks for sharing media

Media sharing networks make it possible to search for and share media on the internet. A photo gallery, a video gallery, and a live video section are available [44]. This service allows you to access and share images, movies, videos, and other files over the internet. Instagram, Snapchat, and YouTube are just a few examples.

Forums for discussion

Among the earliest forms of social media are discussion forums, which offer an ideal platform for market research [45]. They provide information and discussion about various topics. News, information, and viewpoints can be found, discussed, and exchanged using the service. Reddit, Quora, and Digg are a few examples.

Networks for bookmarking and content curation

Users can research and discuss current events and media [46]. For people who are willing for new ideas and knowledge, these platforms represent the epicentre of innovation.

Uses: Find out about new and trending information and media, save it, trade it, and discuss it.

Networks of consumer reviews

Users can share their thoughts and experiences about products, services, brands, locations, and anything else on consumer review networks [47].

Uses: To learn about businesses, restaurants, products, services, travel destinations, and other topics by researching, reviewing, and sharing information and opinions. Yelp, Zomato, and TripAdvisor are among the examples.

Social networks that are not identifiable

As the name implies, these social networks allow users to post material [48] anonymously. As a result, criminals are increasingly turning to such forums to engage in cyberbullying. They use it to stay anonymous when listening in on conversations, people expressing themselves, chatting, and abusing others. Some examples are Whisper, Ask.fm, and After School.

Networks for blogging and publishing

Thanks to blogging/publishing networks, we have a platform for posting online content that enables simple discovery, debate, and sharing thanks to blogging/publishing networks [49]. A publishing platform can be a blog like Blogger or a microblog like Tumblr. It can be as interactive as Medium. Use the internet to post, explore, and comment on the material. WordPress, Tumblr, and Medium are notable ones.

Networks of sharing economy

A "collaborative economy network" is another term for it. People use these networks to advertise [50][51], find, share, trade, buy, and sell things and services over the Internet. It is a way to search for, advertise, share, and trade goods and services online. Just a few examples include Airbnb, Uber, and Task Rabbit.

12.2.3 Social media crimes

In this section, we will give detail about the various types of social media crime. Here we concentrate on six types: phishing, photographic editing, scams in online dating, cyber bullying, and baiting with links. Facebook, for example, has around 270 million illegitimate accounts even though social media sites have strong laws. Information from compromised profiles has been accessed, manipulated, and exploited by hackers for various malicious purposes [52]. On such platforms, stalking, bullying, slander, and the broadcast of illegal or pornographic material are all examples of unethical behaviour. Some examples of social media crimes are listed below. Figure 12.6 depict the types of social media crimes.

Phishing

The most compromised social networking site is Facebook. If you cannot log into your account this is due to someone breaking into it and gaining complete control [53]. When it comes to social media hacking, it usually happens as follows: it is making use of easy-to-guess passwords or passwords

Figure 12.6 Types of social media crimes.

used across various platforms; when utilising a public computer, people do not always log out of their account; sometimes passwords are shared inadvertently; social engineering has been used; and it is possible to hack into an individual's login email address.

Photographic editing

An image shape can be easily transformed into another using photo morphing. According to available data, the number of photos shared daily on social media sites is around 3.2 billion [54] [55]. Criminals will have no trouble downloading and misusing media because it is widely available on social networking platforms. Famous figures' photos are morphed and uploaded to adult websites [56]. They are used to blackmail them for sexual or financial favours.

Scams involving offers and shopping. Women are notorious for falling for offers and retail frauds on social media platforms. Cybercriminals [57], for instance, use shopping offers to entice customers to click on their links. When the user clicks it, it invites them to forward it to 20 friends to receive the voucher. Although the customer does not receive a voucher, the cybercriminal obtains their data.

Scams in online dating

Fraudsters use fictitious identities and photos in these scams to interact with their victims [58]. At the beginning of their relationship, they send little gifts like cards and flowers and demand immediate financial assistance [59] for things like recharging their phone, so they can call, book flights, etc. They switch to a new platform for further conversation after befriending the victim.

Fraudsters also record video calls or grab screenshots of the victim's computer screen, which they use to blackmail the victim later.

Cyberbullying

Sending or publishing disgusting or embarrassing comments or content online and making threats to conduct violent crimes are all examples of cyberbullying [60]. It entails sending or distributing derogatory or false information about another individual to humiliate and slander them. Imposters spread the fatal Blue Whale and Momo Challenges via social media sites like Facebook and WhatsApp. Many youngsters die due to these challenges, with many more committing suicides as a result of them.

Baiting with links

The victim is sent a link to encourage them to open their account. It redirects the victim to a bogus landing page where they are requested to input their account credentials [61]. The credentials are subsequently given to the cybercriminal, who uses them for the wrong reasons. Cybercriminals use exited images and create link. The exited image contains information like your e-mail account won ten million dollars. The created link is passed to the victim.

The victim promptly clicks the embedded link, which takes him to his Twitter or Facebook account login page. The cybercriminal acquires the password and has to take complete control of the account after the victim submits his or her account information [62].

12.2.4 Digital evidence analysis

Digital forensic analysis is the systematic process of analysing and evaluating electronic stored information (ESI) to define evidence to support or oppose matters in a civil or criminal investigation and judicial process [63] [64]. Electronically stored evidence must be handled with considerable caution regarding the forensic investigation. The evidence is preserved, and no actions are taken during the analysis that alters the ESI. This is why forensic pictures or equipment replicas produce the best legal results rather than using original equipment or sources. Data stored in the cloud can also be utilised as digital proof.

12.2.4.1 Analytical purpose

Identifying important players and locations of electronically stored evidence is the first step in analysing digital evidence. During the identification process, the consumer must be explicitly communicated with. The initial range should be precisely defined wherever possible. However, this is not always the case [65]. The who, what, when, and possible place and reason for the exam are listed here. The following items have to be included in the documentation and communication:

- Examination focus
- The general nature of the issue
- The time frame of the chain of events
- A logical or erased piece of data
- Information leakage
- Keywords.

Let us explain one by one,

Examination focus

Before the digital forensic inspection begins, the action scope must be determined. Who are the key stakeholders and custodians? What is the best potential electronic evidence source to be visited for collection? There are many reasons for this information, including:

- So that necessary evidence that may affect the case is not omitted.
- It allows estimating the cost in advance and adjusting the scope of the case to meet the real needs.

Therefore, the possible sources of evidence determined later have an impact on increasing costs.

The general nature of the issue

Customer lists, deemed valuable intellectual property (IP) [66], are also the subject of many legal conflicts. It is critical to comprehend the whole nature of the situation. Is it a will, a trust, or a business plan?

By understanding the nature of the matter, a forensic examiner determines what data or file types he or she should be looking for and where that data might be found. Team member misconduct, theft of firm information, fraud, and divorce are just a few instances.

The time frame of the chain of events

The time and date of the alleged incident, or a range of dates, will aid in narrowing the scope of the inspection. When it comes to employee misconduct, when was the employee's last time worked? When was the last time an employee utilised or returned a company-owned device?

In this and similar cases, resist the impulse to get into an employee's computer because you will be erasing potential evidence, particularly date and time stamps.

A logical or erased piece of data

Logical data is data that has not been deleted and may be accessed without the need for data recovery or special tools. Determine which types of material, such as Word documents, Excel spreadsheets, Acrobat PDF files, photographs, and emails, should be included in the exam. Depending on the situation, social media platforms[67] such as Snapchat, WhatsApp, Facebook, and YouTube may necessitate investigation.

A shredder is an empty recycle bin or trash container. Hidden footprints can be identified by deleted data and cleared web history. After deleting the data, we cannot be limited just by the specific logical data collection. Sector by sector, the entire device must be forensically imaged. It is an excellent practice to aggressively extract forensic photographs of an employee's device after they depart the organisation.

Forensic examiners have to define a scope of analysis and comprehend where data is housed to deliver accurate and quick results.

Information leakage

The unauthorised transfer of information from within to outside an organisation is data leaking.

Is anything plugged into the computer, such as an external hard drive or another linked device? Make sure that any data shared with another device is disconnected [68]. USB drives, mobile devices, and backup devices are all examples.

The Internet of Things (IoT) is becoming increasingly prevalent in our lives and should also be considered. Wearable technology, company or rental cars, surveillance cameras, and house attendants are all examples.

Keywords

What names, phrases, and words are you using to find relevant data? Contacts, personal email addresses, directly competing projects, and company names are examples.

12.2.5 Overview of the digital forensics environment

Cloud computing, smartphones, tablets, and flash drives are becoming more and more valuable sources of data. It is no longer possible to use the words "digital forensics" and "computer forensics" interchangeably. In general, digital forensics is divided into five categories, as follows.

Computer forensics

Computer forensics is a branch of digital forensics, computer forensics, or computer forensic science examining evidence obtained from computers and other digital storage devices. Data forensics involves identifying, preserving, recovering, analysing, and presenting facts and opinions about digital data.

It is used in both criminal and civil proceedings where computers are involved. In addition to data recovery methodologies [69] and concepts, some standards and practices provide a legal audit trail with a transparent chain of custody.

Mobile device forensics

In mobile device forensics, forensic methods are used to recover digital evidence from mobile devices using forensically sound techniques.

While the term "mobile device" is most commonly associated with cell phones, it also refers to any device with internal memory and communication capabilities, such as PDAs, GPS devices, and tablets.

It has been widely accepted since the 1990s that telephones are used in crime; however, the science of forensic examination of telephones is relatively new.

Personal and business information is stored and communicated through mobile phones. Mobile phones are being used in online (e) commerce. The following factors are driving the demand for mobile device forensics.

Network forensics

In network forensics, an analysis of computer network traffic is monitored and analysed to establish legal proof, gather data, or uncover intrusions.

Digital forensics dealing with network data differs from other branches because it is volatile and dynamic. Network forensics [70] is typically a proactive investigation because data is lost once moved. Network forensics has two primary applications:

- Detecting intruders and keeping an eye on a network for odd traffic.
- As part of criminal investigations, law enforcement may analyze acquired network traffic.

Analysing forensic data

Forensic data analysis (FDA), a subset of digital forensics, and investigates structured data in the context of financial crime. Data structures are data that come from databases or application systems. These techniques are used to identify patterns of fraudulent behaviour.

Communication, office programmes, and mobile devices, on the other hand, provide unstructured data. Unstructured data analysis necessitates the usage of keywords or mapping patterns [71] due to the lack of an overall framework. Computer forensics and mobile device forensics professionals regularly examine unstructured data.

Database forensics

Database forensics is a form of electronic forensics that focuses on databases and the metadata that corresponds with them. Data cached in a server's RAM may also exist, necessitating the employment of live analytic tools.

A forensic analysis of a database has included timestamps that relate to the update time of a record in a relational database that is being inspected and reviewed for correctness to confirm the actions of a database user. It also focuses on discovering transactions in a database or application that can be used to prove illegal behaviour, such as fraud.

12.3 The Research Design Behind Forensics

Social networking sites can be used as a source of valuable evidencc through social media forensics, which is what this study aims to do. This section concentrates more on the research design behind forensics. Figure 12.3 shows the different steps involved in research field.

These tools demonstrate how evidence collection is carried out in various contexts using EnCase Forensic, CacheBack, and Internet Evidence Finder programs. Evaluation and comparison of these forensic techniques are based on sample testing results in a laboratory simulation.

Data requirements

When it comes to collecting data during the proposed process, there are a few required phases of research. The phases are the following,

1. Analysing related literature, comparing similar tools and examining vendor products.
2. When configuring the testing environment, analysing the selected tools, exploring the literature review, and utilising assertions [72].
3. Several similar approaches are being applied in this study to improve existing approaches for evaluating tools.

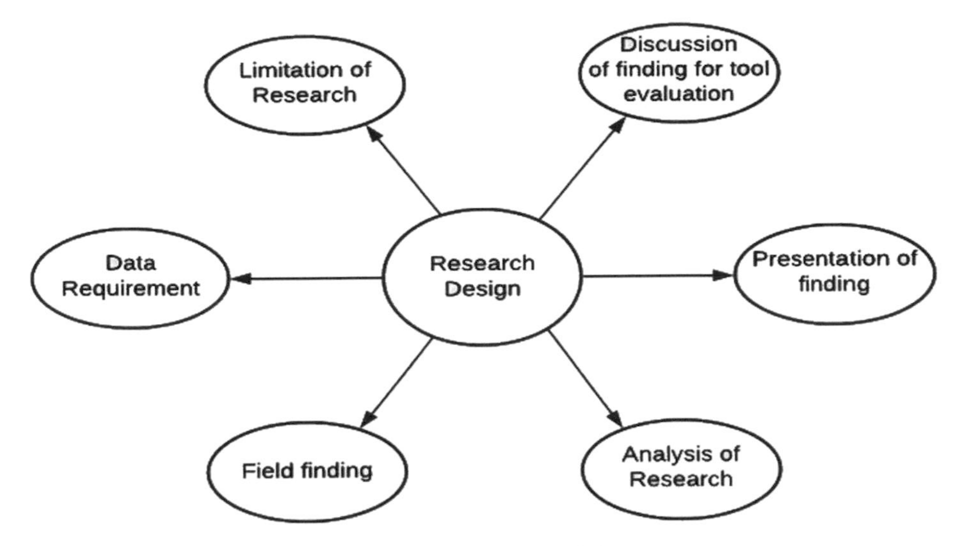

Figure 12.7 Research design.

Digital forensic investigation standards are reviewed against data needs that must be addressed in the research. In the presentation of findings (phase 4), tests are carried out and documented by the testing assessment approach chosen.

The next step is to collect data, where an analytical framework is used to analyse the data appropriately. Analysis of the data is complete after the data has been gathered. Figure 12.7 depicts the research design.

Field finding

There are four phases in the fieldwork.

1. Configuration of testing environments with software and hardware and preparing necessary tools.
2. Validation of gathered evidence and preservation of evidence.
3. Analysis of information collected and evaluation of three tools for evidence extraction
4. Examination of the functionality and results of each extraction tool.

This systematic approach [73] ensures that each phase is completed logically and sequential. Initially, AUT's digital forensics lab is used as a personal research space to set up a thorough testing environment. Phase 3 involves processing the data obtained from the target computer and recording the evidence that has to be identified for each tool examined. Here the phases compared the capabilities of extraction tools against the case requirements as a final step. According to the evaluation criteria and the analysis results from the previous step, the weighed numerical sums for each outcome for each piece of equipment was calculated.

Analysis of research

The accuracy of the data obtained from SNSs is examined at the first stage of data analysis. Analysing the data collected from the social networks is done after the data has been collected. Data from each of the three applications are analysed. As a follow up to the data gathering process, in-depth analysis work will be performed to capture the meaning derived from logical, physical, and professional spaces.

SNS data examination

Investigators examine the social networking data on an external hard drive of a suspect with information separated and blended according to evaluation requirements.

SNS data analysis

Forensic data analysis to determine tool performance and function.

Rating tools

Based on the assessment criteria, weighted numerical totals of the results for each instrument have to be calculated.

Presentation of findings

This part aims to graphically present the findings of the field-testing data from each of the 11 testing situations. Results from three tools, each reflecting their field-testing outcomes, are gathered. One of the tools can examine the underlying data in greater detail. We present the comparative results from the test scenario analysis in the final section.

Discussion of findings of tools

Field finding tests were conducted according to the four phases [74] above to evaluate the effectiveness of the extraction equipment. The three evidence extraction techniques chosen for field testing during the research phase demonstrated the levels of analysis. The results demonstrate that each of the three techniques evaluated in this study can extract evidence from social media sites; however, this power is restricted to extracting pieces of information posted on SNSs.

Creating appropriate conditions to test all three extraction methods is the first step to collecting test data as a baseline. The Windows 7 operating system was installed as a target computer on an existing desktop PC. This investigation requires establishing a stable testing environment and a baseline for the expected evidence since a known baseline is the control.

Quantitative testing is used in the second and third rounds of the testing process. Each tool's test data are collected once the test is completed. A scan of the target machine's hard drive is accomplished using the various scan functions enabled by each tool. Testing can be divided into four stages based on the number of retrieved evidence compared to the expected number of values. Comparative testing is the fourth stage of testing. There is a scale for evaluating the data points of each tool[75].

Based on the results of the data analysis, the capabilities of the three tools are established. Even though specific tools performed much better and extracted significantly more evidence than others, it is reasonable to conclude that evidence from SNSs is not always extractable using these three approaches.

Limitations

When social media sites are exploited in a crime, a digital forensic investigator must determine ways for collecting and analysing evidence from technological equipment. This study uses the capability tests to compare the three technologies and assess the competency of an evidence extraction tool in the context of a social network investigation. Because each instrument has its unique set of capabilities and limits, the proposed research technique includes a number of flaws. In order to perform a forensically sound inquiry and grasp the capabilities of the devices, it is necessary to recognise such restrictions.

Extraction tool evaluations are subject to a number of constraints that must be considered to create reasonable expectations for each tool's performance. The proposed study's first flaw is that none of the three technologies employed is appropriately developed for social network research. The tools utilised in this analysis have several limitations. The three tools were chosen based on their ease of use and market awareness. Although there are various tools for SNS forensics (both open source and commercial software), the study focuses on three well-known and generally available technologies.

The second constraint is that the tool evaluation technique does not establish the complete functionality of each tool because the testing findings are dependent on the research design model presented in this study. Another restriction of the proposed study is that the assessment technique will not be used to examine capacities in all SNS inquiry processes because each one is unique. Furthermore, the testing environment in which the testing equipment is set up has been specifically built. AUT's digital forensic laboratory, constructed specifically for this purpose, has been set up for testing. As a result, it may or may not be entirely representative of the digital forensic investigation environments that examiners would encounter in their job.

12.4 Crime Investigation/Terror Network Structure

Forensic investigations require an understanding of criminal and terror network structures. The analysis of these structures includes determining the rank of people, their affiliations and duties, collaboration among individuals and different networks, spatial analysis based on movement, and network similarity. Figure 12.8 depicts the crime investigation of terror network structures.

Despite the importance of such analysis, current digital forensics tools (both commercial and open source) are incapable of doing a complete analysis. On the other hand, various research projects are undertaken to answer this requirement on a case-by-case basis, focusing on specific events and situations.

The research on network structures and other factors that influence crime and terror networks is divided into categories. Tools and procedures

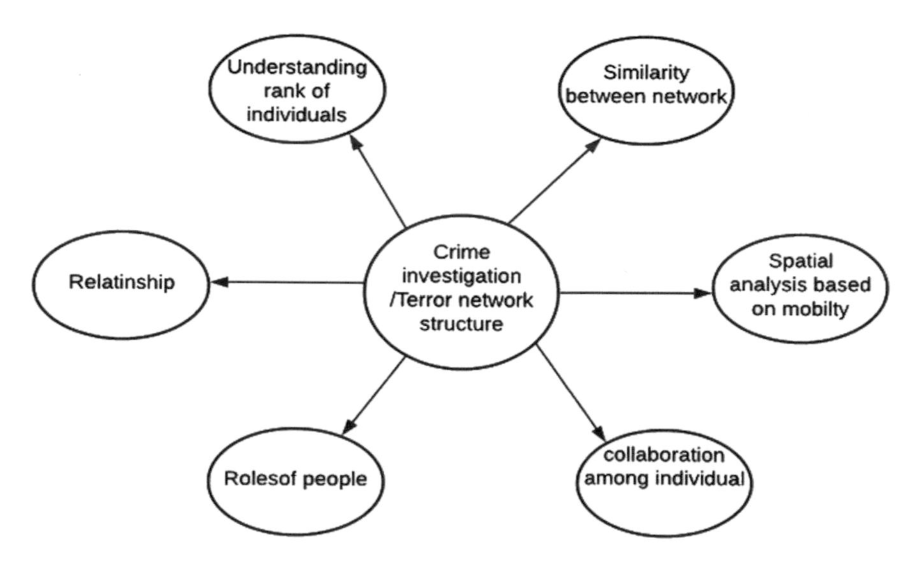

Figure 12.8 Crime investigation.

for additional investigation[76] are necessary once the structures have been identified. As a result, this section will look at the many approaches, tools, and tactics for analysing network topologies, ranking people in criminal networks, measuring network similarities, and looking into diffusion data on these networks.

12.5 Mobile Forensics

Mobile platforms present forensic challenges as data is stored and accessed across multiple devices. Since the data is volatile and can be quickly transformed or deleted remotely, preserving it requires more effort.

Examining and analysing digital evidence from mobile phones is challenging due to their dynamic nature. Technological advancements and innovations are continuously advancing mobile phones. Adapting a single process or tool to examine all devices has been difficult due to the rapid increase in mobile phone types from different manufacturers. Additionally, each mobile phone has its embedded operating system, so forensic experts require special knowledge and skills for acquiring and analysing the devices.

Challenges in mobile forensics

Forensic examiners face different challenges when handling mobile forensics compared to computer forensics. Mobile platforms pose forensic challenges because data can be accessed, stored, and synced across multiple devices. Keeping this data intact requires more effort because it is volatile and can be easily transformed or deleted remotely.

Digital evidence on mobile devices is often difficult to obtain for law enforcement and forensic examiners. The market is overloaded with several mobile phone models from various manufacturers. Different varieties of mobile phone may be encountered by forensic examiners, each with its own size, hardware, features, and operating system. In addition, new models often appear because of the short product development cycle[77]. As the mobile landscape evolves, it is more important than ever for examiners to adapt to all of the problems and stay current on mobile device forensic procedures across various devices.

Security mechanisms are implemented into modern mobile platforms to secure user data and privacy. These characteristics operate as a stumbling block during forensic collection and evaluation. Modern mobile devices, for example, include built-in encryption methods from the hardware to the software layers. The examiner may need to break through various encryption techniques to retrieve data from the devices.

One of the essential forensic requirements is to ensure that data on the device is not altered. To put it in another way, every attempt to extract data from the device should not change the data in it. However, with mobile phones, this is impossible because simply turning on the gadget changes its data. Even if a device appears to be turned off, background processes [78] may continue. Most mobile phones, for example, have an alarm clock that works even when the phone is turned off. Data loss or modification can occur rapidly when switching from one state to another.

Data masking, distortion, fabrication, and secure wiping are anti-forensic tactics that make digital media investigations more difficult.

Unknowingly, digital evidence can be easily tampered with. Browsing an app on the phone, for example, could change the data stored by that app on the device.

Mobile phones have capabilities that allow you to reset everything. Data may be lost if the device is unintentionally reset while being examined.

Moving programme data, renaming files, and changing the manufacturer's operating system are feasible ways to change devices. In this scenario, the suspect's knowledge and experience should be considered.

If the device is password-protected, the forensic examiner must unlock it without destroying the data. While there are ways to get around the screen lock, they may not work in all cases.

Infrared and Bluetooth networks are used by mobile devices, cellular networks, and wireless networks. Because device communication has the potential to modify device data, the prospect of further contact should be ruled out once the device has been seized.

Mobile devices are widely available, but not all tools are accessible. Choosing the right tool for each device can be challenging. Some devices or functions cannot be handled by one tool, so a combination of tools is needed.

12.6 Conclusion

Online social networks (OSNs) allow individuals to freely express themselves online, without constraints on what they want to say or do. This has nurtured a generation of "free spirits" who speak their minds without considering the consequences. Furthermore, OSNs provide a vast audience to whom information can be quickly distributed. According to our research, many tools may extract evidence from social networking sites. However, the evidence is confined to specific fragments of the material posted on SNSs. Because data stored on SNSs is volatile and does not always remain on the user's hard drive, recovering complete data from SNSs is not achievable using the current strategies reviewed in this study.

The findings of this study can be valuable to digital forensic and law enforcement investigators in identifying existing concerns and limitations of the tools and software developers who recognise the rules of the tools to enhance their mining skills

References

[1] Ami-Narh, J. T., & Williams, P. A. (2008). Digital Forensics and the Legal System: A Dilemma of our Times. Australian Digital Forensics Conference (p. 41).ECUPublications.Retrievedfrom https://www.researchgate.net/publication/49280065_Digital_Forensics_and_the_Legal_System_A_Dilemma_of_our_Times AccessData. (2011).

[2] Forensic Toolkit (FTK) Computer Forensics Software. Retrieved July, 2011, from http://accessdata.com/products/computer-forensics/ftk.

[3] Awan, I. (2017). Cyber-Extremism: Isis and the Power of Social Media. Society, 54(2), 138–149. doi:10.1007/s12115-017-0114-0.

[4] Bright, D. A., Hughes, C. E., & Chalmers, J. (2012). Illuminating dark networks: a social network analysis of an Australian drug trafficking syndicate. Crime, Law and Social Change, 57(2), 151–176. doi:10.1007/s10611-011-9336-z. Chaffey, D. (2019).

[5] Global social media research summary 2019. Retrieved from www.smartinsights.com:https://www.smartinsights.com/social-media-marketing/socialmedia-strategy/new-global-social-media-research/.

[6] Chen, L., Xu, L., Yuan, X., & Shashidhar, N. (2015). Digital forensics in social networks and the cloud: Process, approaches, methods, tools, and challenges. International Conference on Computing, Networking and Communications (ICNC) (pp. 1132–1136). Garden Grove: IEEE. doi:10.1109/ICCNC.2015.7069509.

[7] Ahn, Y.-Y., Han, S., Kwak, H., Moon, S., & Jeong, H. (2007). Analysis of topological characteristics of huge online social networking services. Paper presented at the Proceedings of the 16th international conference on World Wide Web Banff, Alberta, Canada.

[8] Anklam, P. (2007). Net work: A practical guide to creating and sustaining networks at work and in the world. Boston: Elsevier/Butterworth Heinemann.

[9] Bascand, G. (2010). Household Use of Information and Communication Technology: 2009. Wellington: Statistics New Zealand.

[10] Bassett, R., Bass, L., & O'Brien, P. (2006). Computer Forensics: An Essential Ingredient for Cyber Security. Journal of Information Science and Technology,3(1), 26.

[11] Beebe, N. L., & Clark, J. G. (2005). A hierarchical, objectives-based framework for the digital investigations process. Digital Investigation, 2(2), 147–167. doi:10.1016/j.diin.2005.04.002.

[12] Berghel, H. (2003). The discipline of Internet forensics. Commun. ACM, 46(8), 15–20. doi: 10.1145/859670.859687 Berghel, H. (2008). BRAP Forensics [Article]. Communications of the ACM, 51(6), 15–20.

[13] Chen, S., Xu, J., Nakka, N., Kalbarczyk, Z., & Iyer, R. K. (2005). Defeating Memory Corruption Attacks via Pointer Taintedness Detection. International Conference on Dependable Systems and Networks (DSN'05) (pp. 378–387). Illinois: IEEE. doi:10.1109/DSN.2005.36.

[14] Glass, G. V. (1976, November). Primary, Secondary, and Meta-Analysis of Research. Educational Researcher, 5(10), 3–8. Retrieved from http://www.jstor.org/stable/1174772.

[15] Hox, J. J., & Boeije, H. R. (2005). Data Collection, Primary vs. Secondary . Encyclopedia of Social Measurement, 593–599.

[16] Huber, M., Mulazzani, M., Leithner, M., Schrittwieser, S., Wondracek, G., & Weippl, E. (2011). Social snapshots: digital forensics for online social networks. Proceedings of the 27th Annual Computer Security Applications Conference (pp. 113–122). Orlando: ACM.

[17] Javed, A., Burnap, P., & Rana, O. (2019, May). Prediction of drive-by download attacks on Twitter. Information Processing & Management, 56(3), 1133–1145. doi:https://doi.org/10.1016/j.ipm.2018.02.003.

[18] Kemp,S.(2019,January2019).Digital2019:GlobalInternetUseAccelerates. Retrieved October 30, 2019, from we are social: https://wearesocial.com/blog/2019/01/digital-2019- global-internet-use-accelerates.

[19] Krishna, N., Fischer, B. A., Miller, M., Register-Brown, K., Patchan, K., & Hackman, A. (2013). The role of social media networks in psychotic disorders: a case report. General Hospital Psychiatry, 35(5), 576.e1–576.e2. doi:10.1016/j.genhosppsych.2012.10.006.

[20] Kumar, S., & Shah, N. (2018). False Information on Web and Social Media: A Survey. Retrieved from https://arxiv.org: https://arxiv.org/abs/1804.08559.

[21] Latonero, M. (2011). Human Trafficking Online: The Role of Social Networking Sites and Online Classifieds. Los Angeles: Center on Communication Leadership & Policy. doi:10.2139/ssrn.2045851.

[22] Lillis, D., Becker, B. A., O'Sullivan, T., & Scanlon, M. (2016). Current challenges and future research areas for digital forensic investigation.

arXiv preprint arXiv:1604.03850, 1–11. Retrieved from https://arxiv.org/pdf/1604.03850.pdf.

[23] Steel, E. (2010, March 29). Nestlé Takes a Beating on Social-Media Sites. The Wall Street Journal, p. B5.

[24] Stewart, D. W., & Kamins, M. A. (1993). Secondary Research: Information sources and methods (2nd ed.). London: SAGE Publications.

[25] Sunstrum, N. (2019). Hacked: A Case Study. Retrieved November 1, 2019, from Michigan Social: https://socialmedia.umich.edu/blog/hacked/.

[26] Tsikerdekis, M., & Zeadally, S. (2014, September). Online Deception in Social Media. Retrieved from uknowledge.uky.edu: https://uknowledge.uky.edu/cgi/viewcontent.cgi?article=1013&context=slis_facpub.

[27] Whittaker, E., & Kowalski, R. M. (2014). Cyberbullying Via Social Media. Journal of School Violence, 14(1), 11–29. doi:10.1080/15388220.2014.949377.

[28] Yusoff, M. N., Dehghantanha, A., & Mahmod, R. (2017). Forensic Investigation of Social Media and Instant Messaging Services in Firefox OS: Facebook, Twitter, Google+, Telegram, OpenWapp and Line as Case Studies.

[29] Contemporary Digital Forensic Investigations of Cloud And Mobile Applications, 41–62. Retrieved from https://arxiv.org/ftp/arxiv/papers/1706/1706.08062.pdf.

[30] Zaharia, A. (2019, May 13). 300+ Terrifying Cybercrime and Cybersecurity Statistics & Trends [2019 EDITION]. Retrieved October 30, 2019, from Comparitech: https://www.comparitech.com/vpn/cybersecurity-cyber-crime-statistics-facts-trends/.

[31] Radack, S. "Forensic techniques: helping organizations improve their responses to information security incidents" 2009. Retrieved September 18, 2015, from http://www.itl.nist.gov/lab/bulletns/bltnsep06.htm.

[32] Parsons, A. "Windows 10 Forensics: Conclusion" - Computer & Digital Forensics Blog, 2015, April 30. Retrieved June 22, 2015, from http://computerforensicsblog.champlain.edu/2015/04/30/windows-10-forensics-conclusion/.

[33] Shavers, B." Virtual Forensics (A Discussion of Virtual Machine Related to Forensic Analysis)", 2008. Retrieved August, 15, 2010. EaseUS. Retrieved June 21, 2015, from http://www.easeus.com/.

[34] Facebook: Monthly active users 2015 — Statistic. (n.d.). Retrieved June 21, 2015, from http://www.statista.com/statistics/264810/number-of-monthly-active-facebook-users-worldwide/ Viber: Number of registered users 2015 — Statistic. (n.d.). Retrieved June 21, 2015, from http://www.statista.com/statistics/316414/vibermessenger-registered-users/.

[35] Doyle, J." A Facebook crime every 40 minutes: From killings to grooming as 12,300 cases are linked to the site", 2012, June 5. Retrieved June 21, 2015, from http://www.dailymail.co.uk/news/article-2154624/A-Facebook-crime-40-minutes-12-300-cases-linked-site.html.

[36] Poh, M. (n.d.). "10 Most Bizarre Crimes Linked to Facebook". Retrieved June 21, 2015, from http://www.hongkiat.com/blog/bizarre-facebookcrimes/.

[37] Vibhuti Narayan Singh, Shalini and G.Khan, "Forensic Analysis of Messaging App Artifacts from Smartphones for Law Enforcement Perspectives",Published in AjMS, Volume 3, Issue 2, Feb 2015, Impact Factor 0.92.

[38] Al Mutawa, N., Al Awadhi, I., Baggili, I., & Marrington, A. " Forensic artifacts of Facebook's instant messaging service", Internet Technology and Secured Transactions (ICITST), 2011 International Conference for (pp. 771–776). IEEE.

[39] Al Mutawa, N., Baggili, I., & Marrington, A. " Forensic analysis of social networking applications on mobile devices", 2012, Digital Investigation, 9, S24–S33.

[40] Baca, M., Cosic, J., & Cosic, Z. "Forensic analysis of social networks (case study"), Information Technology Interfaces (ITI), Proceedings of the ITI 2013 35th International Conference on (pp. 219–223). IEEE.

[41] Mahajan, A., Dahiya, M. S., & Sanghvi, H. P. "Forensic analysis of instant messenger applications on android devices", 2013, arXiv preprint arXiv:1304.4915.

[42] Thakur, N. S. "Forensic analysis of WhatsApp on Android smartphones", University of New Orleans Theses and Dissertations.

[43] Al-Saleh, Mohammed I., and Yahya A. Forihat."Skype forensics in android devices", International Journal of Computer Applications 78.7 (2013): 38–44.

[44] Josh Brunty, " Microsoft Windows 8: A Forensic First Look". (n.d.). Retrieved June 22, 2015, from http://www.forensicmag.com/articles/2012/09/microsoft-windows-8- forensic-fifirst-look.

[45] Nikhalesh Singh Bhadoria, "Windows 8 Forensics Analysis Database" [Tutorial]. (n.d.). Retrieved June 22, 2015, from http://blog.hackersonlineclub.com/2014/01/windows-8-forensicsanalysis-database.html.

[46] Most popular global mobile messenger apps 2015 Statistic. (n.d.). Retrieved June 22, 2015, from http://www.statista.com/statistics/258749/most-popular-global-mobilemessenger-apps/.

[47] Daniel Walnycky, Ibrahim Baggili, Andrew Marrington, Jason Moore, & Frank Breitinger, "Network and device forensic analysis of Android

social-messaging applications". Published in Digital Investigation Impact Factor: 0.99 DOI: 10.1016/j.diin.2015.05.009.

[48] NIST, S. 800–86. "Guide to Integrating Forensic Techniques into Incident Response", 2006, 800–86.

[49] Anglano, C." Forensic analysis of WhatsApp Messenger on Android smartphones". Digital Investigation, 2014, 11(3), 201–213.

[50] Powell, A., & Haynes, C. (2020). Social Media Data in Digital Forensics Investigations. In Digital Forensic Education (pp. 281–303). Springer, Cham.

[51] Kemp, Simon. (2019). 'Digital 2019: Q3 Global Digital Statshot'. Available at: https://datareportal.com/reports/ digital-2019-q3-global-digital-statshot (Accessed: 12th December 2019).

[52] Nikolaidou, A., Lazaridis, M., Semertzidis, T., Axenopoulos, A., & Daras, P. Forensic Analysis of Heterogeneous Social Media Data.

[53] Yusoff, M. N., Dehghantanha, A., & Mahmod, R. (2017). Forensic investigation of social media and instant messaging services in Firefox OS: Facebook, Twitter, Google+, Telegram, OpenWapp, and Line as case studies. In Contemporary Digital Forensic Investigations Of Cloud And Mobile Applications (pp. 41–62). Syngress.

[54] BBC. (2017). 'Cybercrime and fraud scale revealed in annual figures'. Available at: https://www.bbc.co.uk /news/uk-38675683 (Accessed: 12th December 2019).

[55] Montasari, R., & Hill, R. (2019, January). Next-Generation Digital Forensics: Challenges and Future Paradigms. In 2019 IEEE 12th International Conference on Global Security, Safety and Sustainability (ICGS3) (pp. 205–212). IEEE.

[56] Arshad, H., Jantan, A., & Omolara, E. (2019). Evidence collection and forensics on social networks: Research challenges and directions. Digital Investigation.

[57] Soltani, S., & Seno, S. A. H. (2017, October). A survey on digital evidence collection and analysis. In 2017 7th International Conference on Computer and Knowledge Engineering (ICCKE) (pp. 247–253). IEEE.

[58] Caviglione, L., Wendzel, S., & Mazurczyk, W. (2017). The future of digital forensics: Challenges and the road ahead. IEEE Security & Privacy, 15(6), 12–17.

[59] Chaffey, D. (2019). Global social media research summary 2019. Retrieved from www.smartinsights.com: https://www.smartinsights.com/social-media-marketing/social media-strategy/new-global-social-media-research/.

[60] Chen, L., Xu, L., Yuan, X., & Shashidhar, N. (2015). Digital forensics in social networks and the cloud: Process, approaches, methods, tools,

and challenges. International Conference on Computing, Networking and Communications (ICNC) (pp. 1132–1136). Garden Grove: IEEE. doi:10.1109/ICCNC.2015.7069509.

[61] Chen, S., Xu, J., Nakka, N., Kalbarczyk, Z., & Iyer, R. K. (2005). Defeating Memory Corruption Attacks via Pointer Taintedness Detection. International Conference on Dependable Systems and Networks (DSN'05) (pp. 378–387). Illinois: IEEE. doi:10.1109/DSN.2005.36.

[62] Glass, G. V. (1976, November). Primary, Secondary, and Meta-Analysis of Research. Educational Researcher, 5(10), 3–8. Retrieved from http://www.jstor.org/stable/1174772 .

[63] Hox, J. J., & Boeije, H. R. (2005). Data Collection, Primary vs. Secondary . Encyclopedia of Social Measurement, 593–599.

[64] Huber, M., Mulazzani, M., Leithner, M., Schrittwieser, S., Wondracek, G., & Weippl, E. (2011). Social snapshots: digital forensics for online social networks. Proceedings of the 27th Annual Computer Security Applications Conference (pp. 113–122). Orlando: ACM.

[65] Jain, A., & Chhabra, G. S. (2014). Anti-Forensics Techniques: An Analytical Review. Seventh International Conference on Contemporary Computing (IC3). Noida: IEEE. doi:10.1109/IC3.2014.6897209.

[66] Javed, A., Burnap, P., & Rana, O. (2019, May). Prediction of drive-by download attacks on Twitter. Information Processing & Management, 56(3), 1133–1145. doi:https://doi.org/10.1016/j.ipm.2018.02.003.

[67] Kemp,S.(2019,January2019).Digital2019:GlobalInternetUseAccelerates. Retrieved October 30, 2019, from we are social: https://wearesocial.com/blog/2019/01/digital-2019- global-internet-use-accelerates.

[68] Krishna, N., Fischer, B. A., Miller, M., Register-Brown, K., Patchan, K., & Hackman, A. (2013). The role of social media networks in psychotic disorders: a case report. General Hospital Psychiatry, 35(5), 576.e1–576.e2. doi:10.1016/j.genhosppsych.2012.10.006.

[69] Kumar, S., & Shah, N. (2018). False Information on Web and Social Media: A Survey. Retrieved from https://arxiv.org: https://arxiv.org/abs/1804.08559.

[70] Latonero, M. (2011). Human Trafficking Online: The Role of Social Networking Sites and Online Classifieds. Los Angeles: Center on Communication Leadership & Policy. doi:10.2139/ssrn.2045851.

[71] Lillis, D., Becker, B. A., O'Sullivan, T., & Scanlon, M. (2016). Current challenges and future research areas for digital forensic investigation. arXiv preprint arXiv:1604.03850, 1–11. Retrieved from https://arxiv.org/pdf/1604.03850.pdf.

[72] Luxton, D. D., June, J. D., & Fairall, J. M. (2012). Social Media and Suicide: A Public Health Perspective. American Journal of Public Health, 102(S2), S195–S200. doi:10.2105/AJPH.2011.300608.

[73] Myhre, J. W., Mehl, M. R., & Glisky, E. L. (2017). Cognitive Benefits of Online Social Networking for Healthy Older Adults. The Journals of Gerontology: Series B, 72(5), 752– 760. doi:10.1093/geronb/gbw025.

[74] Ning, J., Singh, I., Madhyastha, H. V., Krishnamurthy, S. V., Cao, G., & Mohapatra, P. (2014). Secret message sharing using online social media. IEEE Conference on Communications and Network Security (pp. 319–327). San Francisco: IEEE. doi:10.1109/CNS.2014.6997500.

[75] Park, S., Akatyev, N., Jang, Y., Hwang, J., Kim, D., Yu, W., . . . Kim, J. (2018). A comparative study on data protection legislations and government standards to implement Digital Forensic Readiness as mandatory requirement. Digital Investigation, 24(Supplement), S93–S100. doi:10.1016/j.diin.2018.01.012.

[76] Sindhu, K. K., & Meshram, B. B. (2012). Digital Forensics and Cyber Crime Datamining. Journal of Information Security, 3, 196–201. doi:10.4236/jis.2012.33024.

[77] Statista. (2019). Most popular social networks worldwide as of April 2019, ranked by number of active users (in millions). Retrieved from www.statista.com: https://www.statista.com/statistics/272014/global-social-networks-ranked-by-number-ofusers/.

[78] Steel, E. (2010, March 29). Nestlé Takes a Beating on Social-Media Sites. The Wall Street Journal, p. B5.

13

Blockchain-based Privacy Preservation Technique for Digital Forensics Records

S. Durga[*1], Esther Daniel[2], S. Deepakanmani[1] , T. Mary Neeba[3] and Vinayakumar Ravi[4]

[1]Associate Professor, Department of IT, Sri Krishna College of Engineering and Technology, Coimbatore, India
[2]Associate Professor, Department of CSE, Karunya Institute of Technology and Sciences, Coimbatore, India
[3]Assistant Professor, Department of ECE, Karunya Institute of Technology and Sciences,Coimbatore,India
[4]Center for Artificial Intelligence, Prince Mohammad Bin Fahd University, Khobar, Saudi Arabia
Email: durga.sivan@gmail.com

Abstract

Electronic records of forensic evidence are legally admissible in courts all around the world. Electronic forensic evidence is becoming more widely shared among government agencies so that the investigators can receive a complete view of a case's history. However, maintaining an accurate evidence record and avoiding loss of vital information remains a challenge. Hence the adoption of suitable techniques in this field is a major requirement as they will make information available to authorized users promptly and securely. This research aimed to develop and evaluate a blockchain-based technique that addresses the issues of electronic evidence security, user data rights, and data integrity. In the proposed blockchain-based e-evidence record management system, the Ethereum blockchain is used to create a smart contract that contains a public key for a record saved in the decentralized InterPlanetary file system (IPFS). A proxy re-encryption theme is employed to preserve the user's privacy. In addition, we built and deployed smart contracts to manage business logic for the network's member organizations. Furthermore, this chapter proposes an access control policy algorithm for enhancing information accessibility between legal departments. The prototype system has been developed to implement and test the performance of the proposed technique. Police investigators can utilize the system to retrieve the case data, and validate the authorized user's consent to view it, according to the findings. The access log is stored in a visible and unassailable manner in the blockchain, which is used for audits. The proposed blockchain-based system achieves better results in terms of latency, throughput, and average storage time when compared to traditional evidence record maintenance systems that use client-server architecture.

13.1 Introduction

Digital forensics is a division of forensic science that focuses on detecting, collecting, interpreting, evaluating, and reporting electronically stored data. Almost all alleged crimes involve electronic evidence, and digital forensics assistance is critical for law enforcement investigations. Digital forensics plays a major role in incorporating the overall incident response strategy. As a result, the company should address it through its rules, procedures, budgets, and staff. Digital forensics has extended its coverage to a wide range of technologies. Digital forensics is deep-rooted in the cyber security sector that is a major and necessary aspect of an incident response strategy addressing electronic data. In European countries the courts have approved the latest rules for civil procedures to recognize digital information as an acceptable form of evidence and implemented a mandated mechanism, known as electronic discovery (eDiscovery), to provide the basic idea for this new branch of forensic science. The basic goal of digital forensics is to conduct a technological inquiry within a legitimate framework in effect to illicit acts involving electronic devices. To resolve any disputes between the parties involved in the transactions the utilization of reviewed and tested forensic tools by a trained professional is essential. All necessary policies and procedures should be written in such a way that the digital forensics process is as effective as possible.

Blockchain is a decentralized computation and information exchanging platform that enables multiple domains to coordinate and collaborate towards a rational decision-making process. Incorporating blockchain technology for forensics applications will result in substantial benefits towards the process of data collection, analysis, investigation, validation, and reporting. Blockchain technology improves the transparency of every transaction and thus early and accurate identification of the data sources is achieved. This results in low-cost forensics investigations. Blockchain-based technology determines the chain of transactions and the entities involved in the chain that impacts the investigations in digital forensics. Blockchain is implemented by a distributed data structure involving node entities spread across the networks to construct a transaction that allows all participants in a dispersed network of computers to construct a digital ledger for recording and storing every activity and event. The ledgers help in the investigations of audit trails. Cryptographic hashing and encryption techniques in blockchain potentially establish secure access of evidence in the chain of transactions.

By incorporating blockchain into digital forensics the following features are improved [2]:

Data availability and integrity: The backing up of records in several locations increases the data availability and the integrity of the data stored can be independently validated using the blockchain.

Efficiency: Utilizing blockchain to verify integrity is not time-consuming thus efficiency is achieved by swift data modification verification. Faster processing of the data is also attained.

Fraudulence detection: Blockchain-based process automation and backing up several copies of data in various remote locations will enable forensic readiness and reduce the probability of deletion.

Traceability: All the records are hashed as a part of process automation that establishes a chain of blocks. Hence there is no risk of data modification or data loss. Any modifications or mismatches can be easily be identified with the hash values.

Encrypting enables securing the evidence and adding the blockchain capabilities allows only the intended entities to access the records or data. While accessing any record or data the complete transaction history like time, date, UserID, and system ID will be recorded and appended to the blockchain records through the smart contracts. This makes the records unalterable and they have to be decoded to be utilized by legal nodes. This study focuses on blockchain-enabled forensic investigations that can give self-verification of digital evidence using hash functions to establish verifiability of the evidence. The rest of the chapter is organized as follows: Section 13.2 discusses the background details of the employed technology. Section 13.3 outlines the related work. The proposed blockchain based technique is explained in Section 13.4. Section 13.5 discusses the implementation and performance analysis. Section 13.6 completes the chapter a conclusion.

13.2 Background

Blockchain is a distributed ledger technology (DLT) with several notable characteristics, including a decentralised and trustworthy ledger of records [1]. Blockchain is a chain of blocks that are linked together and are constantly

Figure 13.1 Blocks in a blockchain.

growing as transactions are stored on the blocks. This platform employs a redistributed approach to data transmission, in which each bit of dispersed information, or knowledge, is held in common ownership. Peer-to-peer networks administer blockchains, which hold batches of transactions that are hashed for security. Without any third-party influence, a blockchain has bound edges such as security, privacy, and integrity of the information. These edges provide a cost-effective way to store data, and as a result of technological advancements in the aid industry, the security of sensitive data has become a top responsibility. A variety of researchers have conjointly known that victimization blockchain technology in aid would be a possible resolution. The blockchain is a decentralised application made up of a sequence of blocks linked together in an enormous peer-to-peer network. The headers of the blocks contain the hashes of previous blocks. A block consists of three elements: the square measure of knowledge, the present block's hash, and the prior block's hash. The "genesis block" is the first block in a blockchain, and its previous hash is 0 because there is no preceding block. Figure 13.1 shows the blocks in a blockchain.

Because it is dependent on the type of blockchain, the information could be anything. The information is made up of electronic money coins, similar to bitcoin [1]. These blocks' hash comprises a secure hashing algorithm (SHA) 256 cryptographic rule that is used to uniquely identify each block on the chain. Features of blockchain are described as follows.

- Decentralization: Instead of being stored in a single location, blockchain data is spread across the network. This also allows for knowledge management to be dispersed and handled by consensus obtained through shared input from the network's nodes. Information that was formerly concentrated on a single purpose is now managed by a number of reliable organisations.
- Data transparency: In any technology, having a trust-based connection between entities is required to achieve information transparency. The data

or record in question must be secure and tamper-proof. Any data stored on the blockchain is spread across the network rather than being concentrated in a single location and controlled by a single node. Information is now shared, making it clear and safe from third-party interference.

- Security and privacy: Blockchain technology employs field of study functions to provide security to the network's nodes. To ensure knowledge integrity, SHA-256 was employed to compute the hashes that are retained on the blocks. Blockchain is a suburbanized platform constructed securely by scientific discipline methodologies, making it an honest choice for private information.

13.3 Literature review

The world has shifted to a digital domain, which has resulted in rapid technological innovations and advancements. However, as a result of this transformation, cybercrime and security breach instances have emerged, affecting users' privacy and security. Because of privacy and security, digital forensics is playing a major role, which includes social, Internet of Things (IoT) and cloud forensics. Recently, these technologies aid cyber security professionals in identifying hackers by analysing the digital data generated by data processing and storage. However, specific challenges to digital forensics have been identified in the research, comprising operational, practical, and personnel-related difficulties. The tremendous difficulty of these developed systems, the large data, the chain of custody, personnel security, and the reliability and correctness of digital forensics are all significant barriers to their widespread use [3].

For obtaining and distributing evidence, as well as managing digital forensic evidence, an IoT framework for digital forensics [4] was developed utilising blockchain. Throughout the investigative process, the framework employs novel consortium blockchain to handle a secure chain of custody using case-chain. This framework was designed by using the integration of lattice-based encryption with a new hash function which develops post-quantum resistance and is less complex. The blockchain-based case-chain offers a clear method for all digital investigation and their internal operations. Latency, throughput, gas consumption, energy, and resource use were all considered when evaluating the framework. The goal of this work is to address the issue of investigation accessibility in the field of forensics of digital data.

New blockchain infrastructure is developed to provide comprehensive forensic services for accident investigations [5]. To provide membership establishment and privacy, initially a vehicular public-key infrastructure (VPKI) was introduced into the proposed blockchain infrastructure. They established a fragmented ledger to record extensive data about a vehicle, such as storing the history, car diagnostic information, and so on. Furthermore, the usage of aliases for identities helps users to maintain their privacy. The technique of maintaining and documenting the chronological history of handling digital evidence is known as a digital forensic chain of custody (CoC) [5]. CoC is critical in any digital forensic investigation since it preserves every minute detail about forensic data as it travels through various levels of hierarchy, such as from the first responder to higher authorities in charge of cybercrime investigations. CoC keeps track of details including how evidence was acquired, analysed, and maintained for production, as well as when, where, and who touched the evidence. Lone et al. proposed forensic-chain: a blockchain-based digital forensics chain of custody, carrying truth and tamper resistance to digital forensics chain of custody [7]. Forensic chains have the potential to provide significant benefits to forensic applications in particular and audit trials in general by ensuring the integrity, transparency, authenticity, security, and auditability of digital evidence and operational procedures used during the investigation to achieve the desired outcome.

An alternative to a centralized network for the custody and distribution of digital files in forensic medicine field was developed [8]. This framework was initiated by examining the primary concerns from the patient's standpoint, as well as the perspectives of authorities and applicable laws, and the current, most prevalent official guidelines. However, they all appear to be inadequate at times when it comes to maintaining confidentiality, health professional obligations, and computer forensics difficulties. As a result, they offered a hybrid platform that uses a consensus approach to keeping track of access history and prevents data alteration by unauthorized access. The network is secure and can be accessed via a specific application. To avoid single points of failure, all information is decided upon and distributed across blockchain nodes, and safe access to files is ensured through the use of cryptography and the blockchain key agreement. As a result, a secure and comprehensive framework for uploading, storing, and sharing digital forensic data has been created and implemented.

A novel and secure Hyperledger Sawtooth enabled blockchain architecture for multimedia digital forensic investigations was developed [9]. While gathering, maintaining, analysing, and understanding digital evidence, the proposed MF-Ledger architecture enables comprehensive data provenance, traceability, and assurance for executing various activities as well as confidence between the chain of custody events. The proposed architecture ensures that the entire transactional evidence, which is stored in distributed blocks of data in an encrypted ledger, is protected from intrusion. The proposed architecture enables case-related investors to request, access, and store their digital transactions on the distributed ledger using a blockchain DApp. The private permission blockchain technology Hyperledger Sawtooth was chosen because it has a modular structure that differentiates the decentralized core system from the application domain [11, 12].

Table 13.1 summarizes the core concept and limitations of state of art approaches.

Table 13.1 Core concept and limitations of state of art approaches.

State-of-the-art techniques	Core concept	Limitations
Ethereum blockchain-based digital forensics chain of custody[10]	Digital forensics chain of custody using Ethereum based Hyperledger architecture	Not assured on digital evidence integrity, authenticity, monitoring platform
Financial crime investigation using blockchain-based forensic model [14]	Digital forensics framework, which is an integration of standardized evidence flow and chain of custody techniques	Not able to handle all challenges of financial crime investigation
Digital forensics using blockchain: chain of custody [15]	Analyses the authorized collection of digital evidence, distributed chain of custody to preserve the information	Not able to handle digital crimes fully Able to address only individual crimes
LEchain: lawful proof evidence method for digital forensics [13]	The management model can monitor the evidence collection low, court votes, trail-based outcomes	Information accessibility and auditing are not completely supported Limited security Forensics privacy issues are not addressed
Digital forensics chain of custody using Hyperledger composer [7]	Blockchain-based digital forensics architecture	There is no guarantee of evidence preservation security Poor evidence integrity
The chain of custody using distributed ledger blockchain for modern digital forensics [10]	Distributed ledger for evidence Implements chronologically ordered documentation	Control over data is less

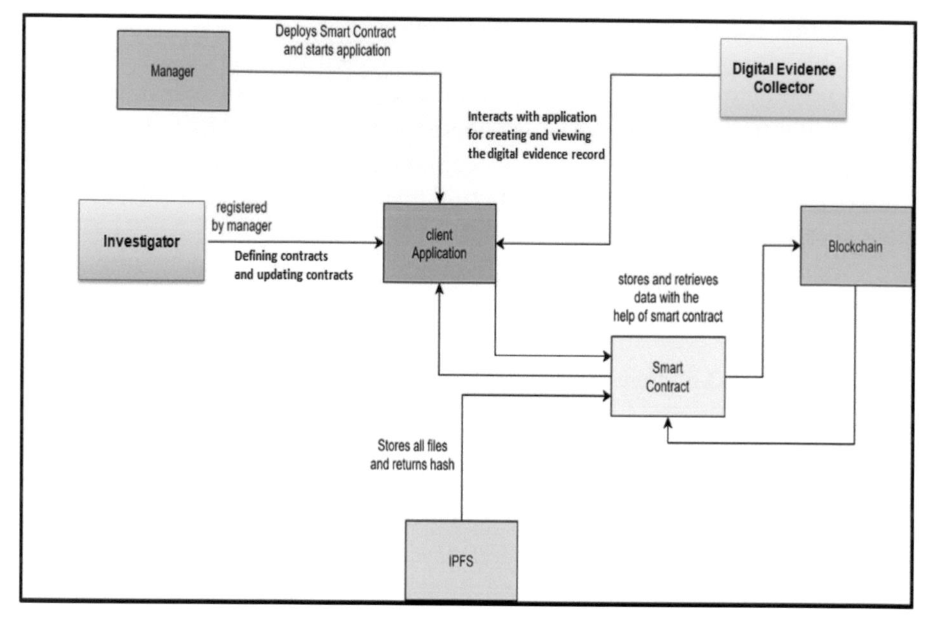

Figure 13.2 Block diagram of a blockchain-based privacy preservation technique.

13.4 Blockchain-based Privacy Preservation Technique

Figure 13.2 depicts a block diagram of the proposed blockchain-based privacy preservation technique for digital forensics records. As shown in Figure 13.2, this allows the system to construct a distributed ledger for documenting digital evidence such as suspects' e-mail or mobile phone files, fingerprints, and other types of digital evidence. All the authorized evidence collectors will share this evidence via the blockchain network. The integrity, time stamping, resilience, provenance, and global trustworthiness of evidence are all guaranteed by block chain's cryptographic nature.

The block diagram consists following critical components:

Consensus Algorithm

The consensus protocol is a critical component of any blockchain network, since it allows all of the network's peers to agree on the current state of the distributed ledger [10]. The power of work (PoW) consensus mechanism is used by Bitcoin, a decentralised digital currency, to determine which node will mine the next block[12].

Ethereum

Ethereum is a blockchain platform that has its own currency, Ether (ETH), as well as a programming language, Solidity. Ethereum is a decentralized public ledger that may be used to verify and record transactions as a blockchain network.

13.4.1 Information transaction

The way an outside unit interacts with Ethereum is referred to as a transaction. It will allow external users to amend the status of records or data held on the Ethereum blockchain network. An activity begun from the outside account is referred to as an Ethereum transaction. Ethereum transactions [16] comprise the following elements: from, to, price, information, gas, and gas value. The first two elements "from and to" represent the sender of the message and the recipient of the message. The amount of money (wei) sent from the sender to the receiver is referred to as the price. The message to be conveyed to the recipient is stored in the optional field named as "information". Each transaction on the Ethereum blockchain has a fee associated with it that the sender must pay. Gas is the name given to this cost. Gas is a cost that the sender must pay for each transaction on the Ethereum network. Gas is the name for this charge. Each transaction's gas value is the amount the sender is willing to spend on gas. The maximum amount of gas that can be obtained in these transactions is called the gas limit.

The distributed ledger system can provide a complete overview of evidence items, with ties to related evidence items and sources. This will come in handy in a range of investigations involving a large number of evidence sources and actions. In the suggested technique, the blockchain is leveraged to give global confidence to all forensic investigation participants.

13.4.2 Smart contract life cycle, user roles, and permissions

A smart contract is a piece of software that is used to conduct any operation on the blockchain. Once the users send the transactions, this piece of code is no longer active [16]. They immediately run on the blockchain, making them impenetrable to tampering and modifications. Smart contracts are written in the solidity programming language and used to programme any exact function on the blockchain. The programmers will compile the desired operations using EVM byte code, which will be explained in the next section, once they have been designed. And once they've been collected, they might be decommissioned and placed on the Ethereum blockchain [16].

A reasonable contract must first be put on the blockchain before it can process transactions. All blockchain nodes should have the same code, according to the deployment mechanism. Different blockchain types handle this need differently, but there are generally two options: keep the contract

Table 13.2 User roles and permissions.

Role	Permission
Evidence collector	• Read/write on permission evidence record • Request the investigator to give access to the record
Investigator	• Read the evidence record • Permission from Courts/Defence/Prosecutions to read/write the record • Revoke permission
Manager	• Create investigator record

code itself within the blockchain, which ensures worldwide agreement, or allows the node owner to decide whether or not to install the code locally. The hash-based promise within the blockchain has been used as a reference for confirming the code reliability.

In addition to data and timestamps, the technique requires the device's unique identifying information, user identification information, and operation details to demonstrate the primitiveness of data.

The roles and permissions of each entity of the proposed technique is shown in Table 13.2.

In the digital investigation, maintaining the integrity, value, and/or ownership of certain evidence pieces is still an issue. Hacking is the cause of many cases, and a vast number of IoT devices are networked. The primary goal is to ensure the integrity of this evidence. Trusted insider risks are on the rise in many scenarios, and vital evidence information has been lost or compromised as a result of insecure evidence systems. To secure the integrity of specific evidence pieces, SHA256 is the cryptographic hash algorithm commonly used in forensic imaging processes; however, there is currently no continuous integrity check or validation mechanism in place for the full evidence chain.

The participants of the network such as the evidence collector, investigator, and apps first generate an RSA key. They maintain their private key safe and secure while making the public key public. Every participant can have a publicly announced public key and a private key that corresponds to it. The digital content is encrypted using an advanced encryption standard (AES) algorithm. To encrypt this information, the AES bilateral coding rule is utilized and the encryption key has been generated arbitrarily. Further, an IPFS has been used to save encrypted data.

When a record is saved to an IPFS, it generates a unique hash for each new record. Anyone who has the hash will be able to access the information. This is now available to the public on Ethereum.

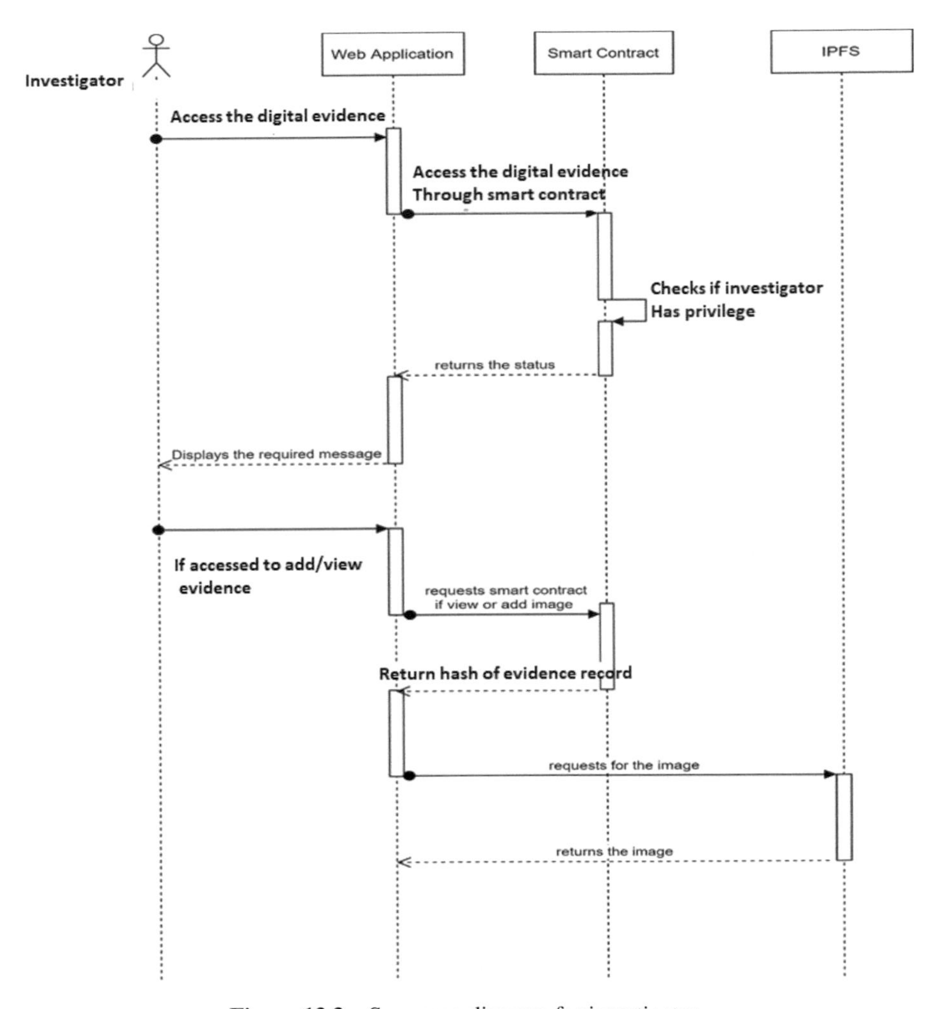

Figure 13.3 Sequence diagram for investigator.

When registering an identity, the authentication server provides the investigator with a private key that belongs to a peer from a national police agency. The sequence diagram for the authorized investigator is shown in Figure 13.3.

The pseudo-code of the proposed technique is as follows.

 # Pseudo code for Investigator Registration

input: address of the sender, image hash, specialization, description
if sender == manager then
    create an investigator instance and store it in the state of contract

```
        push investigator to list
    else
        return error
    end if
```

 # Pseudo code for creating digital forensic record

```
    input: details of the accused person
    if sender not in map of accused person then
        create a new record
        put record to accused person map
    if hash length != 0 then
        set report hash
    end if
    if forensic record hash length != 0 then
    set forensic record hash
    end if
    end if
```

 # Pseudo code on setting report hash

```
    input: access id, hash
    if accessid== investigator id then
        push the hash into the list of report hashes
    else
        return error
    end if
```

Through the provided private key, the investigator obtains the power to access the blockchain and can store the investigation and identity information according to the pseudo-code given. Users without the keys are unable to view or save data on a blockchain. ID, name, social security number, department, jurisdiction, date, investigation information, and evidence ID are among the actions provided to the blockchain by the investigator. Furthermore, only investigators with recognized identities and those working in the legal system have access to the chain code. Following the delivery of the transaction, the smart contract creates a block containing the identity and investigation information, and the transaction and block are transmitted to all blockchain participants. Evidence elements are gathered and then written to the blockchain network, which ensures that each item's authenticity is complete. On the blockchain, all evidence elements are shared among network participants. Without the use of a trusted third party, the proposed technique creates a public time-stamped log for all inspectors. All the evidence is chained and cannot be tampered with.

13.4.3 Interplanetary file system

The IPFS is a peer-to-peer network protocol used for storing and distributing huge files [17]. Because IPFS data is secured from alteration, it provides secure knowledge storage. It uses a cryptanalytic symbol to protect data from tampering, as any effort to modify data saved on IPFS can only be done by changing the symbol dynamically. A cryptographically generated hash price can be found in all IPFS information files. It's distinct, and it's used to distinguish IPFS retain files [17]. The safe storage strategy of the IPFS protocol makes it a viable alternative for storing vital and sensitive data. To reduce total process operations across the blockchain, the resulting cryptographic hash could be maintained on the suburbanized application. The IPFS protocol is based on peer-to-peer networking. The associate link consists of an array of unstructured binary knowledge [17]. The IPFS protocol operates in the following ways:

- IPFS files are assigned a unique cryptanalytic hash.
- On the IPFS network, duplicate files are not permitted.
- A network node stores the node's content and index data.

IPFS is used for storing data locally. It will be able to store and exchange a wide range of files, and it will assign a single hash price to each file, allowing us to easily identify files based on their hash price. Furthermore, an IPFS includes a de-duplication algorithm that can efficiently avoid knowledge storage duplication and conserve storage space. The suggested technique uses blockchain technology, smart contracts, and cryptocurrencies to create the approach to digital forensic applications and services. These apps and services are seamlessly enabled by digital records.

13.5 Performance Analysis

Ethereum, a distributed and open-source blockchain with smart contract capability, was used to evaluate the proposed technique's performance. The evidence collector will choose the investigator to whom view and edit writes are provided while constructing a smart contract (forensic evidence record). The hash of the record stored in IPFS is stored in the blockchain for future reference. Investigators can use smart contracts to obtain the hash contained in the blockchain to get their IPFS records. IPFS stores the investigation details and proofs provided by the investigator.

The implementation of the proposed blockchain-based privacy preservation technique is shown using a sequence diagram in Figure 13.3. For application analysis and testing, Apache JMeter version 5.1.1 and Apache Version two.00 were utilized as desktop performance testing tools. With the addition of more transactions, the execution time will grow. These transactions are carried out to perform the functions that are contained within the sensible contracts. An information payload field is present in every Ethereum transaction. This transaction has a knowledge payload in hex-serialized format with bytes associated with it. Two elements must be present in the knowledge payload: 1. operate Selector, 2. carry out Arguments. The first four bytes of the Keccak-256 hash are used to identify the sensible contract operation that is being invoked by the operate selector. The operation arguments include a variety of static and dynamic component types, each with its own set of criteria for writing in the payload. The operate signature includes the pair of (string, address):

The Keccak-256 hash is as follows,
0x6c0abd24edce8ce20a2dfb1cd2026179214468cde47681e871b6e14bf-9d39efd

The offset in bytes measured from the beginning of the worth coding is supplied explicitly for the static type, but not for the dynamic type. The information payload and block size are used to calculate the dealing size. The payload is of the type String and has a size of 64 bytes. The secret writing of this perform with solely its dynamic varieties is as follows:

0x6c0abd24
0x0000000000000000000000000000000000000000000000000000000000000
00006
0x48976c7c7f2c20667f626c6421000000000000000000000000
00000000000000

Dealing fees in Ethereum are determined in 'ETH' [15], which is the Ethereum coin, as well as its bound units such as dynasties and gwei. The product of gas spent and gas value is the dealing cost for a transaction. It can be written as follows:

Gas consumed × gas price = transaction fee.

The estimated figure for gas consumption is 21000 , which is equivalent to 21 Gwei in terms of gas value. Gwei = 441000 × 21000 = 21000 Gwei.

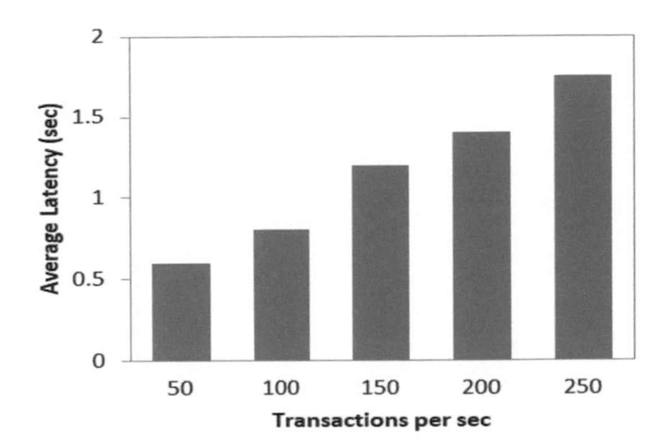

Figure 13.4 Transactions vs average latency.

The following method was used to determine the 1ether dealing fee:

Transaction fee for one Ether = 441000=1000,000,000 Gwei = zero.00041 Gwei

To assess the technique's performance, throughput and average latency were taken into account. Throughput refers to the number of successful transactions per second. Latency is the time it takes for a transaction to be completed. The Ethereum Caliper benchmark results were used to evaluate the performance of the proposed blockchain.

Figure 13.4 shows the effect of changing the transaction delay for write and read operations on the Ethereum platform. A batch of transactions ranging from 50 to 250 is used to analyse the performance. The query latency increases from 0.2 to 1.4 seconds as the number of transactions increases, which are still considered fast. The number of transactions submitted concurrently and the block size, on the other hand, define the average time of an invoke operation. We are using the default block size of ten transactions per block.

The impact of varying the number of transactions on the throughput is shown in Figure 13.5. The blockchain throughput dropped as the transaction rate grew, and the delay considerably increased when the total number of transactions exceeded 200. This means that this particular test setup could handle up to 200 "open" transaction transactions per second without experiencing substantial network latency. When transactions involve fewer operations such as read-only, write-only, not both read and write, a large number of transactions can be supported with substantially reduced latency. Even though the blockchain platform is hosted by a high-capacity hardware system, the type of transactions has an impact on network latency due to the complexity and quantity of operations involved.

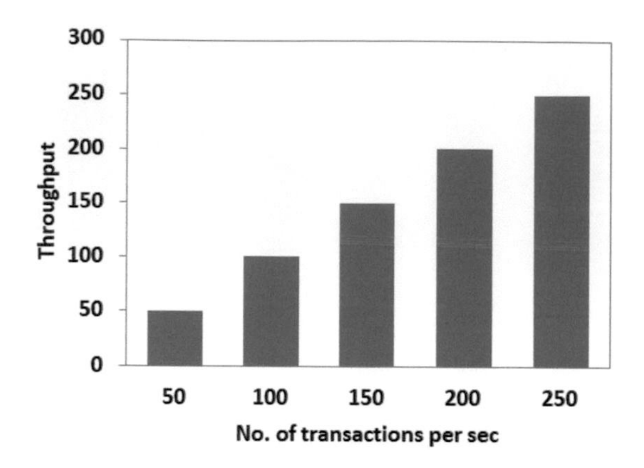

Figure 13.5 Transactions vs throughput.

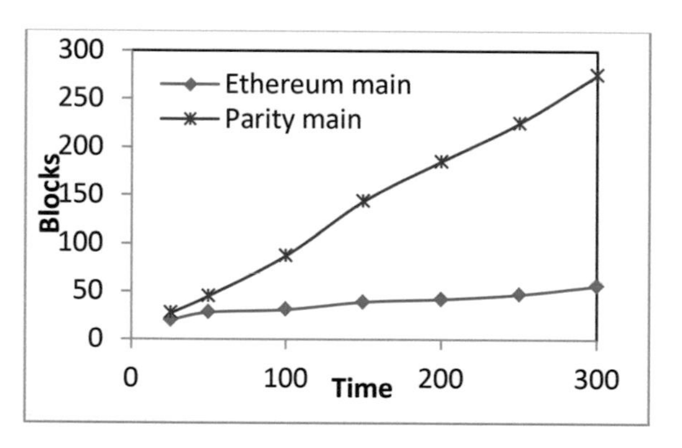

Figure 13.6 Comparison of blockchain forks caused.

The proposed approach focuses primarily on data preservation and security. Tests have been conducted on Fabric versus Ethereum and Parity, as well as on two-dimensional hash versus standard hash table based functions, to see how durable and dependable the model is to crash failures. Figure 13.6 shows the comparison of blockchain forks caused. The forks in the blockchain caused by attacks are seen in Figure 13.6. The impact of the attack causes a linear increase in the number of forks from 50 to 300 seconds. Parity based approaches a fork in the 150th second, as seen in Figure 13.6. Over time, the gap between the number of blocks on the main chain and the overall number of blocks widens. Due to the security of its consensus method, Ethereum, on the other hand, does not have a fork.

Figure 13.7 depicts the effect of the total number of data entries on the average storage time. The input data was chosen at random from a range of

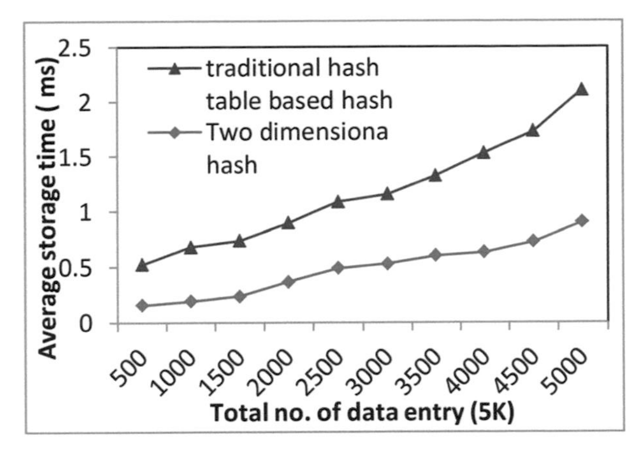

Figure 13.7 Impact of total number of data entries on average storage time.

sizes ranging from 500 to 5000 K. As the size of the data block grows larger, the storage time increases. However, when using a two-dimensional hash, the proposed method takes much less time to store the data blocks. This is due to the fact that retrieving both keys and values at the same time with each keyword is faster.

13.6 Conclusion

This paper examines how a blockchain-based technique might be used to handle concerns such as electronic evidence security, user data rights, and data integrity. In terms of throughput and latency, the effects of the blockchain network workload on the Ethereum platform were investigated. Experiments revealed that the proposed blockchain-based technique for protecting forensic evidence records outperforms the traditional forensic system. To strengthen the digital forensic system, we want to integrate SDN-based network forensics as well as cloud forensics in the future.

References

[1] Vujičić, D., Jagodić, D., & Ranđić, S. (2018, March). Blockchain technology, bitcoin, and Ethereum: A brief overview. In 2018 17th international symposium infoteh-jahorina (infoteh) (pp. 1–6). IEEE.

[2] Al-Khateeb, H., Epiphaniou, G., & Daly, H. (2019). Blockchain for modern digital forensics: The chain-of-custody as a distributed ledger. In Blockchain and Clinical Trial (pp. 149–168). Springer, Cham.

[3] Alghamdi, M. I. (2020). Digital forensics in cyber security-recent trends, threats, and opportunities. *Periodicals of Engineering and Natural Sciences (PEN)*, *8*(3), 1321–1330.
[4] Kumar, G., Saha, R., Lal, C., & Conti, M. (2021). Internet-of-Forensic (IoF): A blockchain based digital forensics framework for IoT applications. *Future Generation Computer Systems*, *120*, 13–25.
[5] Cebe, M., Erdin, E., Akkaya, K., Aksu, H., & Uluagac, S. (2018). Block4forensic: An integrated lightweight blockchain framework for forensics applications of connected vehicles. *IEEE Communications Magazine*, *56*(10), 50–57.
[6] Giova, G. (2011). Improving chain of custody in forensic investigation of electronic digital systems. *International Journal of Computer Science and Network Security*, *11*(1), 1–9.
[7] Lone, Auqib Hamid, and Roohie Naaz Mir. "Forensic-chain: Blockchain based digital forensics chain of custody with PoC in Hyperledger Composer." *Digital investigation* 28 (2019): 44–55.
[8] Lusetti, M., Salsi, L., & Dallatana, A. (2020). A blockchain based solution for the custody of digital files in forensic medicine. *Forensic Science International: Digital Investigation*, *35*, 301017.
[9] Khan, Abdullah Ayub, et al. "MF-ledger: blockchain hyperledger sawtooth-enabled novel and secure multimedia chain of custody forensic investigation architecture." *IEEE Access* 9 (2021): 103637–103650.
[10] A. H. Lone and R. N. Mir, "Forensic-chain: Ethereum blockchain based digital forensics chain of custody," Sci. Practical Cyber Secur. J., vol. 1, no. 2, pp. 21–27, 2017.
[11] H. Al-Khateeb, G. Epiphaniou, and H. Daly, "Blockchain for modern digital forensics: The chain-of-custody as a distributed ledger," in Blockchain and Clinical Trial. Cham, Switzerland: Springer, 2019, pp. 149–168.
[12] S. Bonomi, M. Casini, and C. Ciccotelli, "B-CoC: A blockchain-based chain of custody for evidences management in digital forensics," 2018, arXiv:1807.10359. [Online]. Available: https://arxiv.org/abs/1807.10359
[13] Velliangiri, S., Manoharan, R., Ramachandran, S., & Rajasekar, V. R. (2021). Blockchain Based Privacy Preserving Framework for Emerging 6G Wireless Communications. IEEE Transactions on Industrial Informatics.
[14] L. Zarpala and F. Casino, "A blockchain-based forensic model for financial crime investigation: The embezzlement scenario," 2020, arXiv:2008.07958. [Online]. Available: https://arxiv.org/abs/2008.07958

[15] M. Chopade, S. Khan, U. Shaikh, and R. Pawar, "Digital forensics: Maintaining chain of custody using blockchain," in Proc. 3rd Int. Conf. I-SMAC (IoT Social, Mobile, Anal. Cloud) (I-SMAC), Dec. 2019, pp. 744–747

[16] Yang, R., Wakefield, R., Lyu, S., Jayasuriya, S., Han, F., Yi, X., .& Chen, S. (2020). Public and private blockchain in construction business process and information integration. Automation in construction, 118, 103276.

[17] Velliangiri S., & Naga Rama Devi G.,. (2021). Hybrid Crypto Techniques for Secured Multimedia Big Data Content Protection System (SMBDCPS). International Journal of e-Collaboration (IJeC), 17(2), 1–21. http://doi.org/10.4018/IJeC.2021040101

[18] Li, S., Qin, T., & Min, G. (2019). Blockchain-based digital forensics investigation framework in the Internet of Things and social systems. IEEE Transactions on Computational Social Systems, 6(6), 1433–1441.

14

Multilevel Consensus Blockchain Algorithm for Digital Forensics on Medical Data During the COVID 19 Situation

G. Shanmugarathinam[1], Sheetal[2], Pedro C. Flores[3], and Angelin Gladys[4]

[1]Associate Professor, Presidency University, Bengaluru
[2]Assistant Professor, Presidency College, Bengaluru
[3]Lecturer, Higher College of Technology, Dubai UAE
[4]Lecturer, Department of Information Technology, University of Technology and Applied Sciences, Ibri, Oman

Abstract

This chapter discusses a novel approach to healthcare and the medical field by incorporating blockchain technology. During the COVID 19 situation, the medical field encountered difficulty in managing medical records and gathering accurate data, which seemed to be quite challenging to governments. The proposed method can be used to manage medical data, in particular to connect all governments or private hospitals, securely transfer the data, and focus on digital forensics. Similarly, this proposed method is used to track patients, hospital details such as number of beds, medicine and patient treatment, so that governing bodies can access accurate details by following the necessary steps to serve the people. This method helps in managing medical record data that is more helpful and secure. In the case of any fraud detected in the records, the forensics team can investigate effortlessly. Fraud cannot meddle with the data due to the security provided by a blockchain. A multilevel consensus algorithm is used for adding transactions in the blockchain distributed network.

14.1 Introduction

The healthcare sector is having difficulty managing medical record data and tracing contacts in the present predicament. It is also challenging because of the complexity of its management and the diversity of its stakeholders, especially in underdeveloped countries with poor infrastructure (i.e., no proper procedures to keep track of data, human resources, etc.). Human resources, financing, and medical policy are all factors to consider. One of the most promising, well-known and widely used technologies is blockchain. It was created in 2011 and provides solutions to medical data management issues. During the COVID 19 pandemic, a blockchain played a key role in designing integrated and fair healthcare businesses that rely on improved levels of reliability. The new coronavirus (COVID-19) has upended global economies and ordinary lives, putting enormous strain on healthcare sectors around the world.

In our quest to optimise healthcare resources and allocate biomedical data, the use of the big associated developing technologies – digital forensics on blockchain – has been critical. A transaction or asset is legally immutable after being validated and recorded in a ledger. This has been one of the blockchain's most significant advantages. When one copy or block of data on a device is changed, hundreds of other copies are commonly left unchanged on hundreds of other systems. Changing the data on all decentralised systems would be nearly impossible. Because it is used to link to criminal behaviour directly, digital evidence is crucial in prosecuting cybercrime. During a cybercrime investigation, it is critical to ensure forensic evidence's integrity, reliability, and auditability as it moves through different levels of the chain of custody (CoC) hierarchy. With the advancement of technology, methods may be utilised for data security, improved productivity, and lower costs. The combination of data encryption techniques, hashing, and preservation of evidence using blockchain is the most prevalent strategy for the medical sector.

14.2 Literature Review

Satoshi Nakamoto is a person or group of people credited with inventing blockchain technology in 2008. Satoshi Nakomoto's groundbreaking work inspires us in leveraging blockchain to secure electronic monetary transactions [1].

Simson L. Garfinkel, the author, provided a framework for current forensic research and asserts that, in order to succeed, society must adopt systematic and extendable approaches to information depiction and forensic proceedings. Today, digital forensics is a must-have tool for investigating computer-related crimes and acquiring digital evidence. This research shows that we are amidst a "golden age of digital forensics" [2].

Coronel et al. discussed how digital evidence can be managed at the client's end, including identification, collection, identifiable proof, selection, and analysis procedures. In addition, an examination of how concepts are applied in cyber forensics focused on the client's side [3].

Breitinger et al. discussed two primary purposes. The initial goal is to summarise all available datasets that investigators can use and how to find them. The second goal is to emphasise the need to share datasets so that researchers can recreate the approach to improve the expected outcomes. Preserving the scene, surveying for evidence, documenting the evidence and scene, looking for evidence, and rebuilding the scene are the processes of a criminal investigation. These steps are critical, but preserving and documenting evidence is crucial. First responders, forensic investigators, police officers, court specialists, victims, and suspects are some of the individuals who can work with digital evidence. Each of them has the potential to influence the evidence [4].

CoC is monitored according to four principles, according to Flores and Jhumka:

1. The insider is not permitted to change the original evidence.
2. When original data is produced, a sound rationale must be provided.
3. Regardless of the nature of the arbitration to which the audit trail is submitted, all events should generate the same conclusion.
4. The person in charge of the inquiry must make sure that all of these rules are followed [5].

Prayudi summarised the scope, concerns, and obstacles that the digital chain of custody faces and provides direction on how to deal with CoC. Forensic format, storage, data security assurance, and storage access control are all facets of dealing with the digital chain of custody [6].

Selamat et al. presented the traceability data. They discussed the traces and sources that are linked with the investigative process. Tracing is a technique for determining the source or cause of a problem. It identifies the traces that have been left in evidence. The traces are gathered and stored and, subsequently, they are examined and analysed. This paper proposes a traceability model [7].

A blockchain-based mobile edge computing paradigm has been developed for data exchange and in-home therapy management. For a distributed system, the model framework employs a network. The blockchain is utilised to store patient therapeutic data [8].

Blockchain helps in maintaining the confidentiality of patient data. The therapeutic framework employs blockchain technology to ensure that the data belongs to the patient. Blockchain protects the data's privacy and ownership. The therapeutic metadata is stored on the blockchain. At the same time, the actual data, such as multimedia files, are kept in a distributed or centralised database. There are numerous inventions in today's modern era [9].

Protecting intellectual property is very important; blockchain is a digital ledger that stores intellectual property with security. In this scenario, the blockchain timestamp marking feature comes in handy. Because the work's creator is the person who first accesses the intellectual property document, blockchain is used to authenticate intellectual property ownership. This can be done by tracing back to a specific timestamp in the blockchain. Blockchain is also utilised for transactions involving intellectual property. The parties involved are the person who is using the intellectual property (IP) and the owner of the IP [10].

Beck et al. discussed how a trustworthy distributed storage system is vital in their work for corporate and organisational relationships. The economy relies on trust between individuals and organisations to create and store vital documents. Blockchain technology may be used in a multitude of industries and businesses, including the shipping industry to track cargo, the pharmaceutical industry to track drugs, and the healthcare industry, to name a few. Blockchain would benefit these industries as it allows the safe maintenance of records and each organisation's rightful ownership of assets. As a result of the fourth industrial revolution, the importance and application of blockchain have increased substantially [11].

The focus of this paper's research was on applying this technology in education. Blockchain technology is currently used by academics to track students' achievements and certificates/degrees. The blockchain stores students' interests, learning experiences, and academic information. This aids in combating the degree of fraud that had been a rising problem in the old system. The blockchain system is complicated to manage. Even genuine users cannot alter their data due to the immutability feature. Furthermore, any technical breakdown can result in severe educational concerns. Collecting digital evidence and preserving digital evidence are the two most important factors in crime scene investigations. Each piece of electronic evidence has a physical form that can be recognised with the naked eye. Computers, mobile phones, cameras, CDs, hard discs, and other electronic evidence are examples of electronic evidence. In contrast, digital evidence is evidence recovered or extracted from electronic evidence. A file, an email, a short message, an image, a video, a log, or a text can all be used. Physical and digital crime scene investigations comprise a few essential building elements, such as preserving the crime scene, inspecting for evidence, and documenting the scene. [12]

Digital evidence management and handling is a time-consuming process. Hence, the institution of law enforcement requires a system environment that facilitates the implementation of managing and handling the digital chain of custody to support the management and handling of cybercrime operations. There are flaws in the current system and this can be addressed using blockchain to make it tamper-free [6].

14.3 Problems

1. Medical personnel data compromise: Medical employees in the healthcare industry have access to much sensitive and secret information about their patients, such as credit card numbers and other financial documents. An untrustworthy employee could exploit the patient's information to threaten or blackmail the patient.

2. Another issue in the healthcare sector is lost value in records. Occasionally, some values in particular features will be missing. This is because doctors do not always take all of the essential lab measurements, or the data has been lost.

3. Health professionals and doctors are not technical experts; thus, dataset descriptions and feature names may appear incomprehensible to those unfamiliar with medicine.

14.4 Methodology

We propose a multilevel consensus algorithm for healthcare. Medical data is classified as patient details, treatment details, hospital details, and medicine details. The data is categorised into two levels for the multilevel consensus algorithm. First category: patient and treatment details are more confidential and secure. Hiding content should be at a low level with private, permissioned, and hidden blockchain. The second category should be: medicine and hospital details, public and open blockchain. Both the levels should be protected by POA consensus blockchain for high security.

Process flow: medical data can be classified as public and private data. The authority should control all the information in the network to avoid

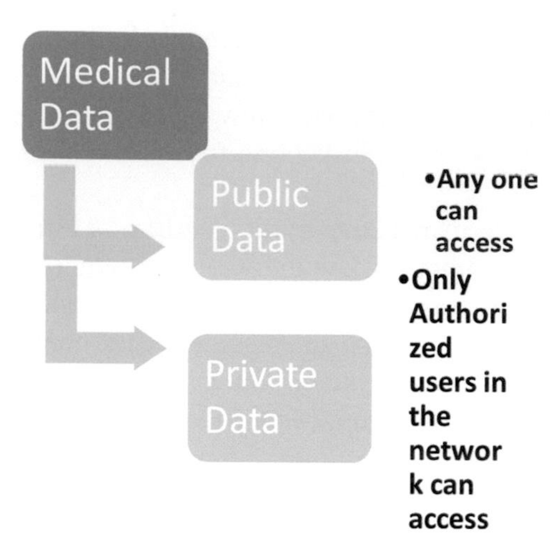

Figure 14.1 Medical data processing based on access and control.

irrelevant information across the network. The required data can be classified further and secured using a blockchain multilevel consensus algorithm to avoid fraud. Any modification as proof cannot be made if the details are stored in the blockchain. Medicine information and the type of treatment given to the patient can also be made available to investigate any complicated case further. Death records to keep track of the patient's reason for death can be proven by using the evidence stored in blockchain technology. Figure 14.1 depict the medical data processing based on access and control model.

Multilevel consensus algorithm: A hybrid consensus mechanism for a hospital management system has been designed, and the layers are as follows:

- Top-level: This should be designed using proof of authority as the primary data to be used and controlled by the top layer users.
- Private and hidden (zero-knowledge protocol, ZKP): The second level is to be used by patients and doctors related to treatment, and other confidential data can be secured and be implemented using zero-knowledge protocol.
- Public and controlled protocol: For emergency-related information regarding beds and medicine availability and the cost of the individual trying to perform the task.
- First category: Patient and treatment details (private and hidden (ZKP)).

  Individual patient and treatment details should be accessed only by people who need the access agreement to do so has to be given by the patient, hospital, and admin staff using a secret key only after all the

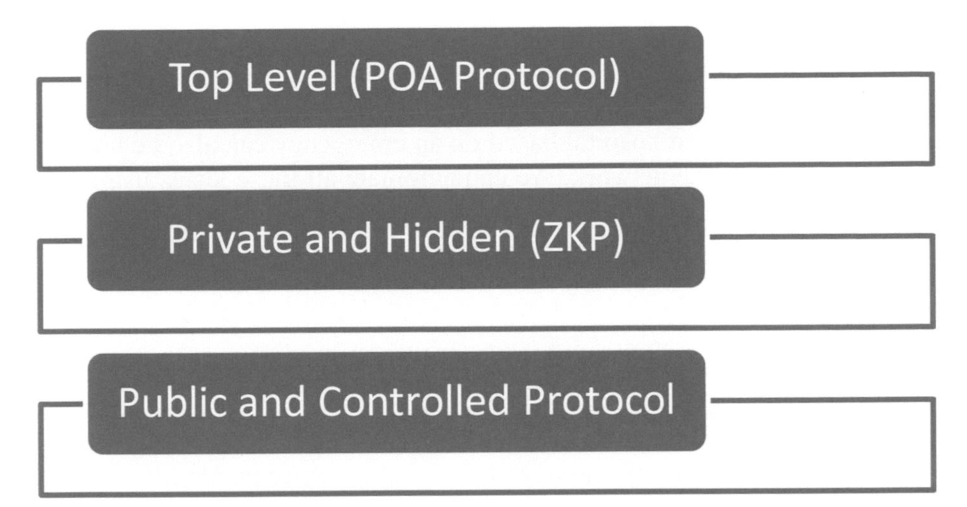

Figure 14.2 Skeleton of the levels.

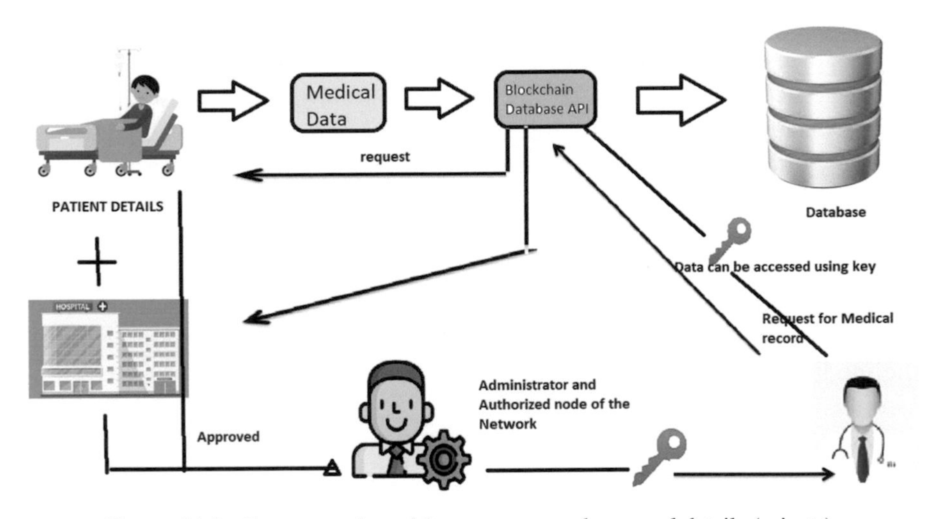

Figure 14.3 Representation of first category and approval details (private).

three stakeholders approve. Treatment details can be accessed by doctors of the hospital enrolled. If the patient wants to go for a second opinion, then the patient can approve the access for other doctors.

- Second category: Medicine and hospital details (public and controlled protocol).

The availability of beds and medicine should be open to everyone to access if they join the public chain of the network created especially for hospitals and blood banks; government agencies can also be part of it. Medicine fraud can also be tracked easily by automated signals, or messages can be generated if there is any problem or alteration in the medicine details available. The nearest hospital based on an emergency can also be notified on request across the network. We can automate all these levels using different consensus algorithms and provide better security and management in the medical field to save the patient. The last layer (top-level) should be protected using a POA, as only certain restricted firms to control the blockchain network.

14.5 Conclusion

This study proposes a new approach for digital evidence management based on blockchain technology. Integrity, transparency, security, authenticity, and auditability are all aspects that blockchain incorporates into its design. As a result, it is one of the most excellent options for maintaining and tracing the forensic chain of evidence. The proposed system will handle the digital evidence from the time it is recovered until submitted as evidence in court. It will ensure the evidence's integrity, traceability, authenticity, and security. It will also aid in the admissibility of this digital evidence in a court of law. It proposes a multilevel consensus algorithm that can be referred to as hybrid consensus to use in the medical field to preserve the evidence of flaws in the medical sector related to staff, doctors, fraud with respective treatment and an easily accessible public chain to save the patient in case of emergency. We have designed a more complicated mechanism that protects the medical data based on the user.

14.6 Further Work

Future work is to build a more effective, more practical consensus mechanism related to blockchain for processing and storing entire evidence files. Medical data to be hidden should be implemented using the zero-knowledge protocol. Implementing multilevel and hybrid is the further work that can be carried out, innovating and proving novel consensus algorithms. This will assist investigators in handling digital evidence carefully, ensuring the digital evidence's integrity. All evidence will be guaranteed to be safe and untampered with in blockchain technology.

References

[1] Satoshi Nakamoto. Bitcoin: A peer-to-peer electronic cash system. Cryptography Mailing list at https://metzdowd.com, 03 2009.

[2] Simson L. Garfinkel. Digital forensics research: The next 10 years. Digital Investigation, 7:S64–S73, August 2010.

[3] Karina Campos Bryan Coronel, Priscila Cedillo and Jessica Camacho. A systematic literature review in cyber forensics: Current trends from the client perspective. In IEEE Third Ecuador Technical Chapters Meeting (ETCM) 2018, pages 1–6. IEEE, 2018.

[4] Frank Breitinger Cinthya Grajeda and Ibrahim Baggili. Availability of datasets for digital forensics and what is missing. Digital Investigation, 22:S94–S105, August 2017.

[5] D. A. Flores and A. Jhumka. Implementing chain of custody requirements in database audit records for forensic purposes. In 2017 IEEE Trustcom/BigDataSE/ICESS, pages 675–682, August 2017.

[6] Yudi Prayudi and Azhari Sn. Digital chain of custody: State of the art. International Journal of Computer Applications, 114:975–8887, April 2015.

[7] S. R. Selamat, R. Yusof, S. Sahib, N. H. Hassan, M. F. Abdollah, and Z. Z. Abidin. Traceability in digital forensic investigation process. In 2011 IEEE Conference on Open Systems, pages 101–106, September 2011.

[8] Abdur Rahman, M. Shamim Hossain, George Loukas, ElhamHassanain, Syed Rahman, Mohammed Alhamid, and Mohsen Guizani. Blockchain-based mobile edge computing framework for secure therapy applications. IEEE Access, PP:1–1, November 2018.

[9] Asaph Azaria, Ariel Ekblaw, Thiago Vieira, and Andrew Lippman. Medrec: Using Blockchain for medical data access and permission management. pages 25–30, 08 2016.

[10] Junyao Wang, Shenling Wang, Guo Junqi, Yanchang Du, Shaochi Cheng, and Xiangyang Li. A summary of research on Blockchain in the field of intellectual property. Procedia Computer Science, 147:191–197, January 2019.

[11] Roman Beck, Michel Avital, Matti Rossi, and Jason Thatcher. Blockchain technology in business and information systems research. Business Information Systems Engineering, 59, November 2017.

[12] Guang Chen, Bing Xu, Manli Lu, and Nian-Shing Chen. Exploring blockchain technology and its potential applications for education. Smart Learning Environments, 5(1), March 2018.

[13] Seokhee Lee, Hyunsang Kim, Sangjin Lee, and Jongin Lim. Digital evidence collection process in integrity and memory information gathering. In First International Workshop on Systematic Approaches to Digital Forensic Engineering (SADFE'05), pages 236–247, Nov 2005.

[14] Auqib Hamid Lone and Roohie Naaz Mir. Forensic-chain: Blockchain-based digital forensics chain of custody with poc in hyperledger composer. Digital Investigation, 28:44–55, March 2019.

[15] Sangjin Lee Hyunji Chung, Jungheum Park. Digital forensic approaches for amazon alexa ecosystem. Digital Investigation, 22:S15–S25, August 2017.

[16] J. Cosic and M. Baca. Do we have full control over integrity in digital evidence life cycle? In Proceedings of the ITI 2010, 32nd International Conference on Information Technology Interfaces, pages 429–434, June 2010.

15

Blockchain-based Identity Management Systems in Digital Forensics

Vani Rajasekar[1], K. Sathya[2], S. Velliangiri[3], and P. Karthikeyan[4]

[1]Assistant Professor, Department of CSE, Kongu Engineering College, Tamilnadu, India
[2]Assistant Professor, Department of CT/UG, Kongu Engineering College, Tamilnadu, India
[3]Assistant Professor, Department of Computational Intelligences Institute of Science and Technology Kattankulathur Main Campus, India
[4]Post-Doctoral Researcher, Department of Computer Science and Information Engineering, National Chung Cheng University, Taiwan
Email: vanikeit@gmail.com; pearlhoods@gmail.com; velliangiris@gmail.com; nrmkarthi@gmail.com

Abstract

The collection, storage, and analysis of big data has become an enormously crucial solution for addressing cybercrime and establishing legal cases since improvements in information advancing technologies during the last two decades. In our interconnected, digitalized society, digital identification is becoming increasingly vital. Identity management solutions are commonly used in real-world applications and are designed to make managing digital identities and activities like verification easier. There have been initiatives in recent times to offer blockchain-based access control solutions that enable individuals to share control of their own identities. A legal technique for collecting, analysing, storing, and reporting digital evidence is known as digital forensics. This type of evidence is critical to police investigations since it depicts the facts of computer hackers' scenes and links suspected to criminal activity. As a result, it's critical to treat obtained evidence with care to make sure that it's acceptable in police investigations and court proceedings, as well as to avoid tampering or wrongdoing. Current studies have turned to blockchain technology to promote transparency, immutability, and auditability for managing such legitimate evidence. In this chapter, we create a concrete blockchain-based authentication and authorization framework. The Hyperledger Composition proof of concept is also supplied, and its efficiency is reviewed. With the range of improvement for a comprehensive end-to-end program, the model has proven reasonable overhead in terms of performance and resource consumption.

15.1 Introduction

15.1.1 Digital identity management (DIM)

Identity management, often referred to as authentication and authorization, is a set of principles and tools that guarantee that the appropriate authorities have access to the appropriate information [1]. Digital identity management (DIM) is defined as a framework model utilized by any computer or communication system to define the identity of users, provide access privileges, and hold accountability. The identity of users plays a vital role in any computer or communication platform. The digital identity (DI) of any user typically has a lifecycle of three phases namely:

- Setup of initial identity
- Maintenance of identity
- Termination of identity.

Digital forensics investigations look into the DI transactions performed in DIM to identify and verify the identity of suspects related to the crime. The identity verified in forensic investigation is acceptable as evidence in a court of law to prove the guilt/innocence of the suspect. DIM is designed carefully with a view to tracing back to the identity of the person responsible for the crime. Digital forensics takes responsibility for identifying the person involved in the crime with proper evidence and no doubt. This chapter explores the various identity management techniques that secure the identity of users with the help of blockchain technology [2]. DIM is said to be forensically ready when the design of DIM aims to minimize the processing time of forensic investigation with little cost. Also, the DIs present in the DIM are represented in a way so that investigation is easier while maintaining the creditability of the evidence.

15.1.2 Digital forensics

Digital forensics is an interesting branch of forensics investigation dealing with digital evidence related to a computer/mobile crime. Processes involved in digital forensics include recovering, investigating, examining and analysing digital content with respect to the crime [3–5]. It is most commonly used in civil or criminal courts to support or refute a hypothesis postulated by any entity. It is also commonly used in corporate or private sectors to ensure the compliance of security policies.

Digital forensics are generally categorized into network forensics, computer forensics, mobile forensics, and forensic data analysis. All these kinds of forensic investigations typically have the following processes:

- Seizure of digital evidence
- Imaging (forensic copying) of evidence
- Analysis of evidence
- Generation of a forensic report.

15.1.2.1 Uses of digital forensics

Digital forensics is useful for various purposes, as listed below:

- Identification of direct evidence in a crime
- Identification of evidence supporting the suspects in a crime
- Confirms statements or alibis of suspects
- Uncovers the intent of a crime
- Reveals the sources of a crime
- Verifies and authenticates the contents of a document related to a crime.

15.1.3 Blockchain technology

Blockchain is a technique used to store digital information in the form of blocks. The blocks are linked together to form a chain in such a way that it is impossible to change the content of a block [6]. This leads to blocks being tamper-proof. The chain of blocks is shared among all the users in the network thus creating a shared database of digital information. The contents of the block are verified and added to the existing chain by the process of mining. The shared database is also known to be distributed ledger.

15.2 Blockchain in DIM

Conventional DIM is insecure as it is maintained in an isolated framework with no relationships. Fragmented DIM is more susceptible to be tampered with by any attacker trying to erase his footprints while committing a digital crime. To ensure the creditability of DIM to be presented as evidence in a court of law, blockchain enabled DIM can be put into use [7-10]. Blockchain helps to store and maintain digital identities securely in an interoperable unified way.

Blockchain based identity management is beneficial due to the following:

- Stamps out inaccessibility problems
- Secures the identity of users
- Prevents or detects fraudulent identities.

Decentralized identifiers (DID) are the most commonly built framework for the digital identity management using blockchain. Blockchain is used in the creation of digital identities for the users with the following modules:

- Decentralized identifier
- Identity management
- Embedded encryption.

15.2.1 Decentralized identifier (DID)

DID is a pseudo-anonymous identifier assigned to an entity (person, company, object, etc.). When an entity registers for creation of a DID, a pair of private and public keys are generated along with the identity. The DID along with public key is stored in the existing blockchain [11]. DIM may choose to store additional information of an entity in the blockchain along with the public key. A private key is never stored in the blockchain to ensure integrity of data. A private key is used by the entity to verify the ownership of the DID. However, there is no limitation on the number of DIDs an entity can hold. A person or entity may hold one or more DIDs across various platforms. Chain of integrity across all the DIDs is achieved by each and every DID attesting for other DIDs. Attestations are made for the conformity of characteristics like age, gender, and ID of the entity [12–13]. The credentials are stored by their owners after getting cryptographically signed by the issuers. Use of public and private key pairs ensures the security of DIDs and ties up closely the identity of user with his/her actions.

15.3 Use Cases of DID

Decentralized identifiers are used in DIMs for one of the following purposes:

- Self-sovereign identity
- Data monetization
- Data portability.

15.3.1 Self-sovereign identity (SSI)

SSI is a framework model that allows entities to store their identity data on their own device and can control what data need to be provided to validators [14]. This system moves the storage of identity data from central server repository to a user's own local device making it completely sealed from outside world.

15.3.2 Data monetization

Entities with DID can trade their identity data for economic benefits, not only for the validation of access control mechanisms, but also for various purposes like machine learning training algorithms, online advertisers, and so on.

15.3.3 Data portability

Data portability allows the entities to port their identity data from one controller to another controller seamlessly. Data owners have the right to choose their controllers, and while making such transitions, the necessity to validate identity data to the new controller is completely eliminated. With DID, switching between controllers is made easy and users can have reusable credentials.

15.4 Benefits of DID

Blockchain based creation of decentralized identifiers have the following benefits:

- Decentralized public key infrastructure
- Decentralized storage
- Manageability and control.

15.4.1 Decentralized public key infrastructure (DPKI)

DPKI is the infrastructure model required by blockchain based DIM to distribute private keys to the identity owners. Private keys are required to be transmitted to the owner in a secure, tamper-proof network to ensure their integrity [15]. When a user creates a DID, the public and private key pairs are generated. The public key is added to the existing blockchain in a tamper-proof way. The private key is held with the owner. Inclusion of public keys into the public blockchain enables validation of identity without the need of any digital certificates from the certification authority.

15.4.2 Decentralized storage

Identities created and hooked into blockchain are safer from being used compared to storing the identity data in a single remote server. In blockchain, identity data are distributed using a DPKI and typically reside on the owner's devices like mobile phones, laptops, and so on [16]. Storage of decentralized identity data is possible with interplanetary file systems. When identities are stores in a decentralized manner, the ability of an entity to gain unauthorized access to identity on a central server is nullified.

15.4.3 Manageability and control

Conventional identity management systems store identity data in a central server. It holds the issuer of identity as responsible for providing the security. In a blockchain enabled identity management system, the user or owner of the identity is held responsible for the security measures. Users may choose their own security mechanism by implementing the necessary security techniques or outsourcing the security needs to third party apps like a digital bank vault, password managing apps and so on.

15.5 Blockchain and Digital Forensics

The technical process of identifying, preserving, collecting, and presenting digital evidence so that it becomes applicable to the case of law is known as digital forensics. In reality, any information stored or retrieved from digital technology might be considered a piece of electronic evidence that can be analysed throughout a digital forensics' inquiry. Because the goal of any forensic examination is to guarantee that the digital evidence produced is acceptable in court, preserving the integrity of the evidence is a vital criterion that must be maintained all through the process of investigation. The process of establishing and sustaining the temporal history of preserving digital data is known as chain of custody (CoC) [17–19]. To prevent the CoC from being tampered with or destroyed without permission, extreme caution is essential.

The ultimate goal of CoC is to show that suspected information is sufficient for the alleged crimes rather than being placed erroneously. The examination of digital forensics necessitates well-defined methods that adhere to industry standards, applicable regulations, and organizational policies. Although forensic investigators employ a variety of tools and techniques, the investigative process generally entails organization, gathering, conservation, examination, and presentation.

15.5.1 Hyperledger composer

Hyperledger is both a framework and an open-source development toolkit for quickly creating blockchain applications. Hyperledger Composer makes the process of designing and deploying blockchain use cases much easier, cutting development cycles. The major advantage of Hyperledger Composer is that it is completely open source, with a communication and organizational architecture that allows anybody to contribute to the development. It supports reconfigurable blockchain consensus protocol which will ensures that all the transactions are validated and authenticated. Hyperledger Caliper is a multi-phased development tool [20]. The first phase is the preparation phase, during which the test context is set up by installing decentralized applications. The second phase is the tests implementation phase, in which clients are given tasks to do in order to execute specified test cases that can be based on transactions counts or length. The final phase is the performance evaluation phase, which gathers all of the test findings in order to generate a report. Figure 15.1 shows the process in hyperledger composer.

15.5.2 Secure forensic model

Forensic investigation prefers to function in controlled contexts with regulations such as the identification of investigators working on criminal cases. While the Bitcoin and Ethereum networks are built on anonymity, anyone can observe the transactions, but it's practically impossible to figure out who was engaged in them [21]. As a result, Bitcoin and Ethereum are unlikely to be the ideal candidates for crime investigation, specifically digital forensic exploration, which necessitates anonymity and is conducted by legitimate and recognized practitioners of intelligence agencies. Untrusted individuals must not be able to tamper with the digital evidence. Hyperledger Composer is designed to meet expectations for creating an automation process that is both resilient and secure in collecting all of the facts relevant to a specific cyber forensic case's gathering evidence. The

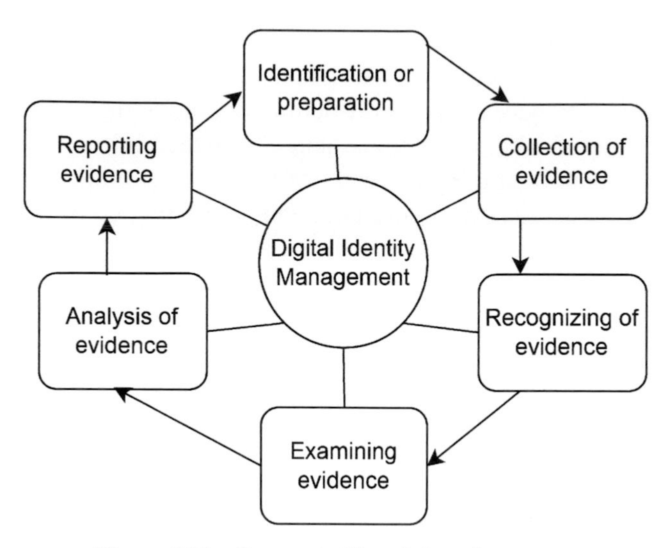

Figure 15.1 Process in Hyperledger Composer.

suggested framework will serve as a foundation for any forensic investigation or record keeping conducted by an organization, allowing them to employ a blockchain-driven methodological framework for an investigation or auditor's report [22]. The proposed forensic architecture consists of four main components namely

- Actors
- Evidence module
- Blockchain network
- Secure storage.

15.5.2.1 Actors

The actors are regarded as the true players in any system that contains the storage of transaction data. Participants typically represent businesses, but they may also represent individuals, authorities, or other interests. Actors in the suggested forensic-chain architecture are forensic investigators, their job is to gather as much information as feasible regarding digital evidence and store it in a blockchain. Prosecutors, defence [23], and courts also participate since they require facts regarding the chain of evidence in the case, which is recorded and maintained using blockchain, at any time during the forensic inquiry. Only authorized participants have access to change the state of the forensic-chain and inspect the specific evidence's information. Figure 15.2 shows the digital forensic architecture.

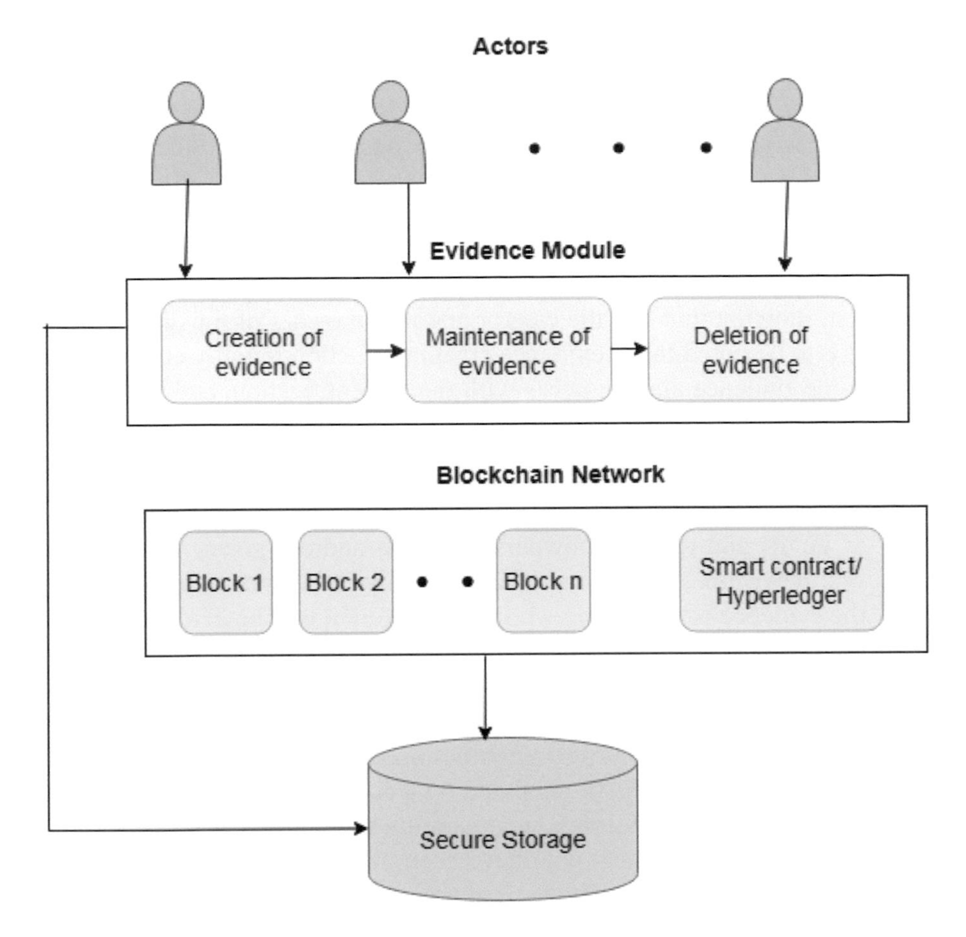

Figure 15.2 Digital forensic architecture.

15.5.2.2 Evidence module

The front-end for the crime scene investigation model is built with Hyperledger Composer. This aids in the creation of REST APIs for the forensic-chain corporate network and the creation of skeletal angular applications for the architecture. Members interact with the forensic-chain through custom-made software [24]. The architecture allows for easier connection with the Ethereum blockchain. Members call a relevant core module to collect and store information elements from the forensic-chain. The evidence module has following sections

- Creation of evidence
- Maintenance of evidence
- Deletion of evidence.

A. Creation of Evidence
The evidence ID and evidence description are entered into the evidence creation module, which then sends the information to the forensic-chain. Because the ID is created by hashing digital evidence, it aids in the preservation and conservation evidence's validity all through the life cycle. Other parameters, such as originator and proprietor, are likewise set to the location of the member who established it for the first time [25]. The address or identifier of the participant is placed in the list, showing that it is the creator and initial owner of the digital proof. It is to be noted in the evidence creation function initially checks to see if the evidence already exists with same identification, and if it does, it exits without producing a copy.

B. Maintenance of Evidence
The evidence transfer function accepts an evidence ID and an address as inputs and transmits ownership to the address given. The function first determines if evidence has been provided, and the proprietor of the evidence is the person who calls the function. The evidence display method accepts an evidence ID as an input and provides evidence data from the blockchain. This function's sole purpose is to ensure that evidence actually existing. The suggested forensic-chain paradigm is based on Hyperledger Composer, a permissioned Bitcoin protocol that runs in a regulated environment regulated by a consortium or single organization. As a result, information concerning the proof is only available to blockchain participants who have been allowed by admin competitors who are controlled by consortium organizations. With the use of channel in the Hyperledger fabric/composer, individuals can transfer information in a secure manner.

C. Deletion of Evidence
The evidence deletion method accepts an evidence ID and erases the evidence associated with it from the forensic-chain. It initially determines whether evidence exists and whether the member who activates it is the evidence's author; if so, the information item is removed from the blockchain. Evidence is a property in the forensic-chain concept; hence the evidence deletion method deletes the related evidence from the resource database. When information is no longer applicable or reliable for the issue under examination, this is necessary. No one can erase actual evidence, but anyone can submit transactions indicating that a shred of information is no longer applicable to a case.

15.5.2.3 Blockchain network

A block chain technology is a technological structure which allows customers to access ledger and cryptographic protocol services [26]. Smart contracts are mainly used to originate operations, which would then be transmitted to each peer node in a network and documented immutably on their own copy of the blockchain.

A. Key Components of Blockchain

- *Smart contract:* A smart contract is a collection of rules that is kept on the network and executed automatically to speed up operations. A smart contract can specify requirements for corporate debt transfers, as well as payment terms for trip insurance [27].
- *Distributed ledger:* The distributed ledger and its unchangeable records of transactions are accessible to all participants in the network. Transactions are now only documented once with this public ledger, avoiding the duplication of work that is common in traditional corporate networks.
- *Immutable records:* After one transaction has been logged to the public ledger, no user can edit or interfere with it. If a mistake is found in a transaction record, a new transaction must be made to correct the error, and both operations then must be accessible.

15.5.2.4 Secure Storage

A blockchain can be utilized in shared storage software in a few various ways.

- Break information down into manageable parts.
- Encrypt the information such that only you have access to it.
- Spread files across a network so that all of your files are accessible, even if a portion of the network is unavailable.

Each fragment is duplicated to prevent unwanted loss in the event of a link failure. The files are additionally encrypted with an encryption key, making it hard for other network nodes to see them. The duplicated fragments are dispersed across the globe by decentralized nodes. The interactions are stored in the blockchain ledger, which allow the network to authenticate and synchronize activities across the blockchain's hubs. The blockchain storage system is meant to keep track of these interactions indefinitely, and the data can never be altered [28]. The huge amount of information that is copied in

Table 15.1 Security requirements.

Security no	Security feature	Description
1	Traceability	Because every communication is recorded in an immutable ledger, actions and evidence can be tracked from their inception to their annihilation.
2	Authentication	To use the cryptography techniques, all individuals and units should be distinct as well as provide conclusive proof of identification.
3	Security	Information can only be added or accessed by actors who have been granted authorization. The operations are authenticated in a cryptographically secure manner due to strong underlying consensus mechanism.
4	Verifiability	The appropriate actors can verify the transactions and connections. This verification is possible to carry out in real time.
5	Non-repudiation	Each activity may be linked to its originator, ensuring a high level of accountability.
6	Integrity	Because hashes are used, the data and evidence from the occurrences cannot be manipulated or corrupted during transport and analysis.

this process results in an overabundance of data. Cloud storage also necessitates enterprise-level technology for its data centres. Because of these characteristics, centralized data storage can be much more costly than blockchain information storage. In addition to the improvement listed above, storing data amongst dozens of unique nodes improves the security of blockchain storage. Hackers will have a harder time accessing the data if the files are encrypted and distributed over the decentralized network. There is no centralized authority in charge of file access or the keys required to decode the data. The user has complete control over the private keys, making it practically difficult for a third party to get access to the information. Sharing further prevents each node from seeing the entire file's contents, resulting in even more confidentiality. The main required security features of digital forensic investigation are specified in Table 15.1.

15.6 Performance Evaluation

The performance of the proposed approach can be analysed with the following parameters.

- Throughput

- Latency
- CPU utilization
- Memory utilization
- Gas.

15.6.1 Throughput

The number of nodes added to the blockchain every second is used to calculate this measure. Throughput of the blockchain is described as the number of transactions per unit time [29]. It is determined by the consensus method in use, which describes how networks interact in order to confirm the legitimacy of the additional transaction as well as the integrity of each of their versions on the public ledger. Figure 15.3 depicts the throughput analysis of different methods.

$$\text{Throughput} = \frac{\text{Total number of transactions}}{\text{Total time for interactive nodes}} \tag{15.1}$$

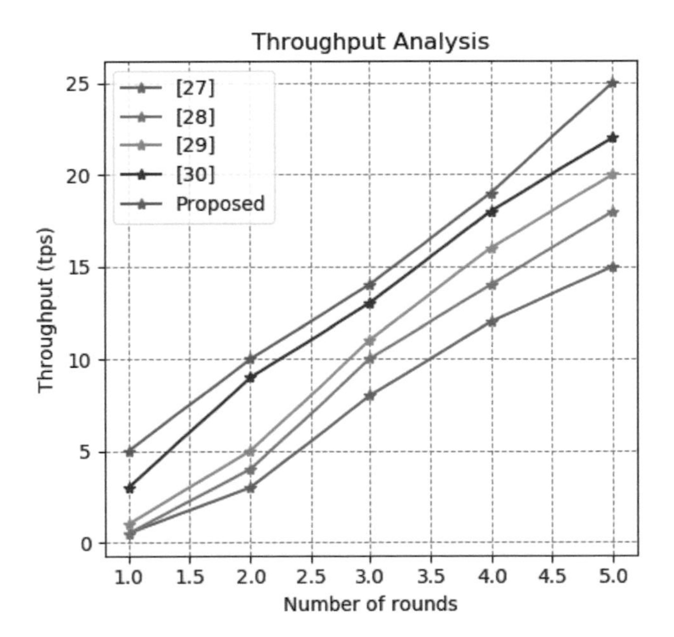

Figure 15.3 Throughput analysis.

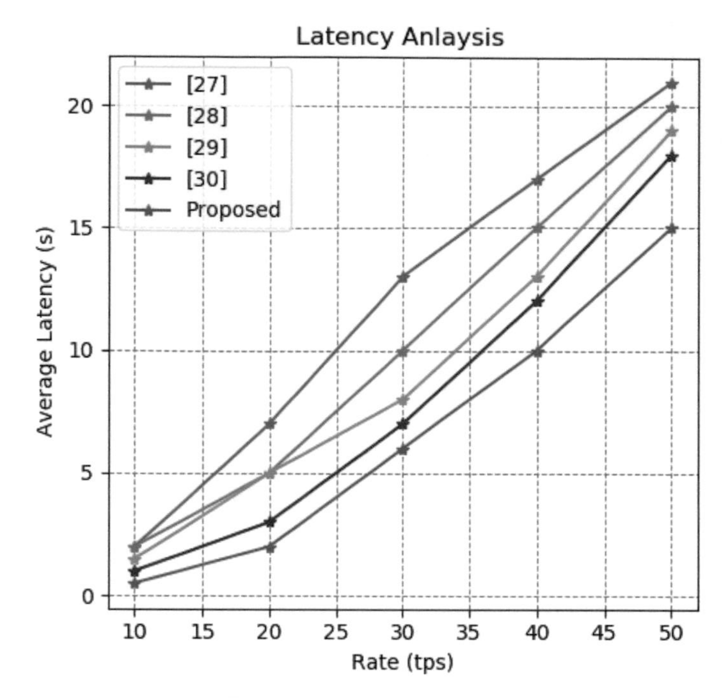

Figure 15.4 Latency analysis.

15.6.2 Latency

The period between the start of a transaction and the transaction being included in the blockchain is known as latency or transactional latency. The total of the block addition and agreement delay is the transaction latency. Figure 15.4 depicts the Latency analysis of different methods.

$$\text{Latency} = \text{Block latency} + \text{consensus time of that block} \qquad (15.2)$$

15.6.3 CPU utilization

The usage of CPU in real time applications provides insight into the service's efficiency. This usage ratio is expressed as a percentage, indicating how much of the CPU processing units is being used to analyse the blockchain in relation to a specific parameter (Figure 15.5).

15.6.4 Memory utilization

The usage of memory in real time applications provides insight into the service's efficiency (Figure 15.6). This usage ratio is expressed as a percentage,

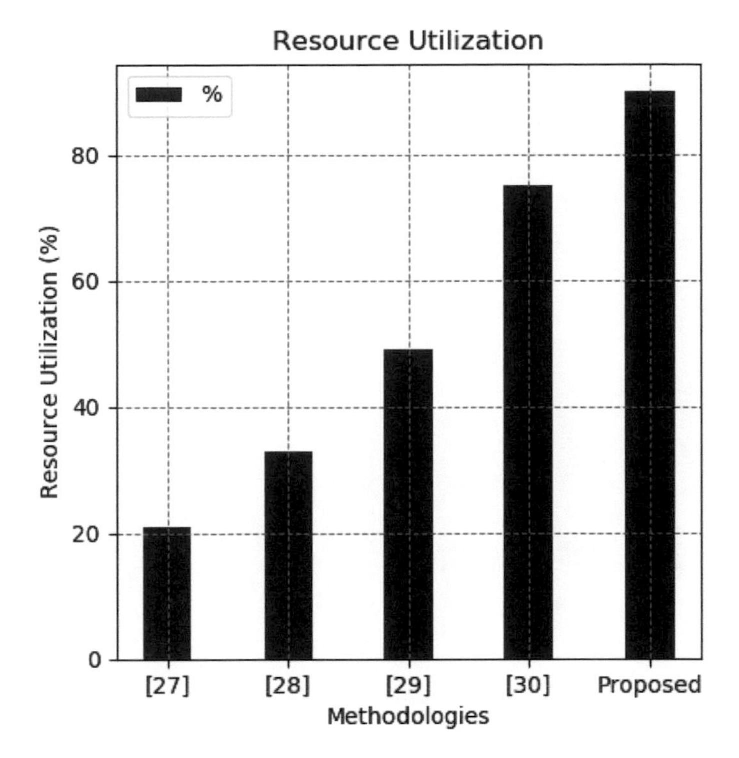

Figure 15.5 Resource utilization.

indicating how much of the system memory is being used to analyse the blockchain in relation to a specific parameter [30].

15.6.5 Gas

Gas is required for each and every transaction or consortium blockchain on the Ethereum network. It is a proportion of an Ethereum token which is used by the agreement to compensate miners for their services in validating the transactions on the blockchain. The amount of gas used in proportion to the size of the frames and the number of transactions.

15.7 Summary

Authenticity, accountability, transparency, scalability, password protection, and secrecy are all implied in blockchain technology. As a result, it demonstrates promise and emerges as a viable choice for improving digital forensics

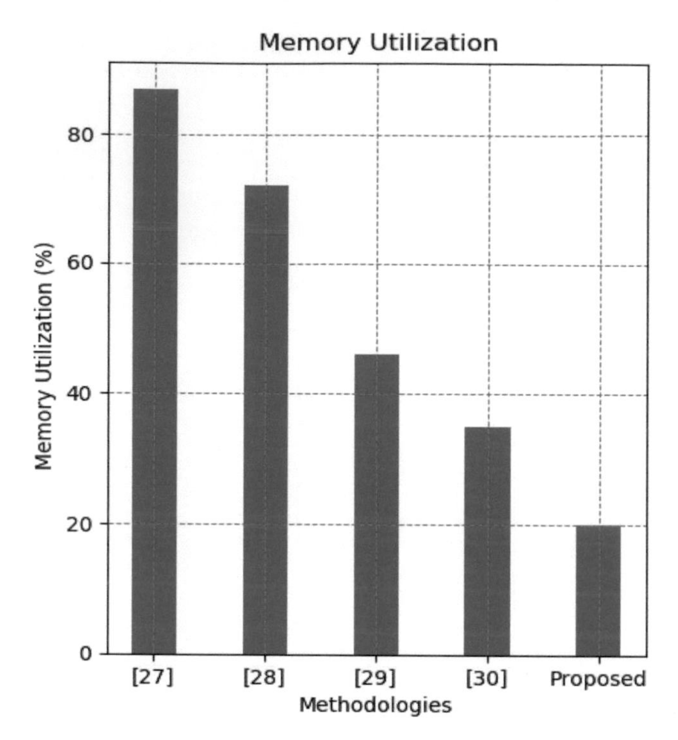

Figure 15.6 Resource utilization.

investigations in the IoT domain. The ever-increasing number of connected devices in the Industry 4.0 presents a threat of an ever-increasing range of attacks and threats. Several study endeavours have been done previously, but each has its own set of constraints. We have provided an answer for digital forensics operations in IoT in this chapter. The usage of blockchain to provide authentication methods in blockchain is a new thought that is tried in the proposed solution. Forensic data is collected using smart contracts. Latency, throughput, gas consumption, energy, and resource usage have all been assessed, and failure points have been executed and analysed. The goal of further work is to create a fully efficient final integrated system for storing digital information and preserving chain of custody, which will be supported by shared data and public blockchain.

References

[1] Kumar, G., Saha, R., Lal, C., & Conti, M. (2021). Internet-of-Forensic (IoF): A blockchain based digital forensics framework for IoT applications. *Future Generation Computer Systems*, *120*, 13–25.

[2] Lone, A. H., & Mir, R. N. (2019). Forensic-chain: Blockchain based digital forensics chain of custody with PoC in Hyperledger Composer. *Digital investigation*, *28*, 44–55.
[3] Zarpala, L., & Casino, F. (2020). A blockchain-based Forensic Model for Financial Crime Investigation: The Embezzlement Scenario. *Available at SSRN 3740042*.
[4] Lin, C., He, D., Huang, X., Khan, M. K., & Choo, K. K. R. (2018). A new transitively closed undirected graph authentication scheme for blockchain-based identity management systems. *IEEE Access*, *6*, 28203–28212.
[5] Zarpala, L., & Casino, F. (2020). A blockchain-based Forensic Model for Financial Crime Investigation: The Embezzlement Scenario. *Available at SSRN 3740042*.
[6] Kevin, D., & David, B. (2018, September). HACIT2: a privacy preserving, region based and blockchain application for dynamic navigation and forensics in VANET. In *International Conference on Ad Hoc Networks* (pp. 225–236). Springer, Cham.
[7] V. S, R. Manoharn, S. Ramachandran and V. R. Rajasekar, "Blockchain Based Privacy Preserving Framework for Emerging 6G Wireless Communications," in IEEE Transactions on Industrial Informatics, doi: 10.1109/TII.2021.3107556.
[8] Zou, R., Lv, X., & Wang, B. (2019). Blockchain-based photo forensics with permissible transformations. *Computers & Security*, *87*, 101567.
[9] Kotsiuba, I., Velykzhanin, A., Biloborodov, O., Skarga-Bandurova, I., Biloborodova, T., Yanovich, Y., & Zhygulin, V. (2018, December). Blockchain evolution: from bitcoin to forensic in smart grids. In *2018 IEEE international conference on big data (big data)* (pp. 3100–3106). IEEE.
[10] Lone, A. H., & Mir, R. N. (2019). Forensic-chain: Blockchain based digital forensics chain of custody with PoC in Hyperledger Composer. *Digital investigation*, *28*, 44–55.
[11] Li, S., Qin, T., & Min, G. (2019). Blockchain-based digital forensics investigation framework in the Internet of Things and social systems. *IEEE Transactions on Computational Social Systems*, *6*(6), 1433–1441.
[12] Kumar, G., Saha, R., Lal, C., & Conti, M. (2021). Internet-of-Forensic (IoF): A blockchain based digital forensics framework for IoT applications. *Future Generation Computer Systems*, *120*, 13–25.
[13] Rajasekar, V., Premalatha, J., Sathya, K., & Saračević, M. (2021). Secure remote user authentication scheme on health care, IoT and cloud applications: A multilayer systematic survey. *Acta Polytechnica Hungarica*, *18*(3), 87–106.

[14] Pourvahab, M., & Ekbatanifard, G. (2019). Digital forensics architecture for evidence collection and provenance preservation in IaaS cloud environment using SDN and blockchain technology. *IEEE Access*, *7*, 153349–153364.

[15] Billard, D., & Bartolomei, B. (2019, June). Digital forensics and privacy-by-design: Example in a blockchain-based dynamic navigation system. In *Annual Privacy Forum* (pp. 151–160). Springer, Cham.

[16] Kumar, M. (2020). Applications of Blockchain in Digital Forensics and Forensics Readiness. In *Blockchain for Cybersecurity and Privacy* (pp. 339–364). CRC Press.

[17] Rajasekar, V., Varadhaganapathy, S., Sathya, K., & Premalatha, J. (2016, March). An efficient lightweight cryptographic scheme of signcryption based on hyperelliptic curve. In *2016 3rd International Conference on Recent Advances in Information Technology (RAIT)* (pp. 394–397). IEEE.

[18] Tian, Z., Li, M., Qiu, M., Sun, Y., & Su, S. (2019). Block-DEF: A secure digital evidence framework using blockchain. *Information Sciences*, *491*, 151–165.

[19] Popescu, A. C., & Farid, H. (2004, May). Statistical tools for digital forensics. In *international workshop on information hiding* (pp. 128–147). Springer, Berlin, Heidelberg.

[20] Baig, Z. A., Szewczyk, P., Valli, C., Rabadia, P., Hannay, P., Chernyshev, M., ... & Peacock, M. (2017). Future challenges for smart cities: Cyber-security and digital forensics. *Digital Investigation*, *22*, 3–13.

[21] Costantini, S., De Gasperis, G., & Olivieri, R. (2019). Digital forensics and investigations meet artificial intelligence. *Annals of Mathematics and Artificial Intelligence*, *86*(1), 193–229.

[22] Elhoseny, M., Selim, M. M., & Shankar, K. (2020). Optimal deep learning based convolution neural network for digital forensics face sketch synthesis in internet of things (IoT). *International Journal of Machine Learning and Cybernetics*, 1-12.

[23] V. S, R. Manoharn, S. Ramachandran and V. R. Rajasekar, "Blockchain Based Privacy Preserving Framework for Emerging 6G Wireless Communications," in *IEEE Transactions on Industrial Informatics*, doi: 10.1109/TII.2021.3107556.

[24] Rajasekar, V., Premalatha, J., & Sathya, K. (2020). Enhanced Biometric Recognition for Secure Authentication Using Iris Preprocessing and Hyperelliptic Curve Cryptography. *Wireless Communications and Mobile Computing, 2020.*

[25] Horsman, G. (2019). Formalising investigative decision making in digital forensics: Proposing the Digital Evidence Reporting and Decision Support (DERDS) framework. Digital Investigation, 28, 146–151.

[26] M. Hossain, Y. Karim, R. Hasan, FIF-IoT: A forensic investigation framework for IoT using a public digital ledger, in: Proc. - 2018 IEEE Int. Congr. Internet Things, ICIOT 2018 - Part 2018 IEEE World Congr. Serv., 2018, pp. 33–40.

[27] Zhihong Tiana, Mohan Lia, Meikang Qiub, Yanbin Suna, Shen Sua, Block-DEF: A secure digital evidence framework using blockchain, Procedia Comput. Sci. (2019) 1–16.

[28] Jung Hyun Ryu, Pradip Kumar Sharma, JeongHoon Jo, Jong Hyuk Park, A blockchain-based decentralized efficient investigation framework for IoT digital forensics, J. Supercomput. 75 (2019) 4372.

[29] J. Ricci, I. Baggili, F. Breitinger, Blockchain-based distributed cloud storage digital forensics: Where's the beef? IEEE Secur. Priv. 17 (1) (2019) 34–42.

[30] S. Rane, A. Dixit, BlockSLaaS: Blockchain assisted secure logging-as-aservice for cloud forensics, in: S. Nandi, D. Jinwala, V. Singh, V. Laxmi, M. Gaur, Faruki P. (Eds.), Security and Privacy, ISEA-ISAP 2019, in: Communications.

Index

F

G

I

K

L

M

N

O

P

R

S

U

V

About the Editors

P. Karthikeyan obtained the Bachelor of Engineering (B.E.) in Computer Science and Engineering from Anna University, Chennai, and Tamil Nadu, India, in 2005 and received his Master of Engineering (M.E) in Computer Science and Engineering from Anna University, Coimbatore India in 2009. He completed his PhD degree at Anna University, Chennai, in 2018. He is currently working as post-doctoral researcher in national Chung Cheng University, Taiwan. He was skilled in developing projects and carrying out research in the area of Cloud computing and Data Science with programming skills in Java, Python, R and C. He published more than 20 International journals with a good impact factor and presented more than 10 International conferences. He was the reviewer of Elsevier, Springer, Inderscience and reputed Scopus indexed journals. He is an editorial board member in EAI Endorsed Transactions on Energy Web, The International Arab Journal of Information Technology and Blue Eyes Intelligence Engineering and Sciences Publication journal.

Hari Mohan Pandey received the B.Tech. Degree from Uttar Pradesh Technical University, India, the M.Tech. Degree from the Narsee Monjee Institute of Management Studies, India, and a PhD degree in computer science and engineering from the Amity University, India. He worked as a Postdoctoral Research Fellow at Middlesex University, London, U.K. He also worked on a European Commission project – Dream4car under H2020. He is a Research associate professor at Bournemouth University, U.K. He is specialized in computer science and engineering, and his research area includes artificial intelligence, soft computing, natural language process, language acquisition, machine learning and deep learning. He is the author of various books in computer science and engineering, published over 50 scientific papers in reputed journals and conferences. He has delivered keynotes, invited lectures and served as session chair for international conferences. He has served as a leading guest editor for several journals such as Swarm And Evolutionary Computation Elsevier, Neural Commutating and Applications Springer, Neural Processing Letter Springer, Intelligent Decision Technology

Top Press, Wseas Transactions On Systems, International Journal Of Artificial Intelligence And Paradigm Inderscience, and International Journal Of Artificial Intelligence and Soft Computing Inderscience. He has served as a reviewer for Applied Soft Computing, Elsevier, Ieee Transactions On Evolutionary Computation, Ieee Transactions On Cybernetics, Acm Transactions On Multimedia Computing, Swarm And Evolutionary Computing Elsevier etc. Dr Pandey is an Associate Fellow of the Higher Education Academy (U.K. Professional Standard Framework) and has rich experience in teaching at the higher education level. He was the recipient of the Global Award For The Best Computer Science Faculty of the year 2015, the award for completing Indo-Us Project Gentle, an award (Certificate of Exceptionalism) from the Prime Minister of India, and an award for developing Innovative Teaching And Learning Models for higher-education.

Velliangiri Sarveshwaran obtained his Bachelor's in Computer Science and Engineering from Anna University, Chennai. Masters in Computer Science and Engineering from Karpagam University, Coimbatore, and Doctor of Philosophy in Information and Communication Engineering from Anna University, Chennai. Currently, he is working as an Assistant Professor at SRM Institute of Science and Technology, Kattankulathur Campus, Chennai. He was a member of the Institute of Electrical and Electronics Engineers (IEEE) and the International Association of Engineers (IAENG). He was the reviewer of IEEE Transactions, Elsevier, Springer, Inderscience, and reputed Scopus indexed journals. He is specialized in Network Security and Optimization techniques. He published in more than 30 International journals and presented at more than 10 International conferences. He also serves as a technical program committee and conference chair at many international conferences. He also serves as an area editor in the EAI Endorsed Journal of Energy web journal.